The Marijuana Industry

INSIDER'S PLAYBOOK

The Marijuana Industry Insider's Playbook

Copyright © 2018 by Stansberry Research. All Rights Reserved.

All rights reserved. No part of this book may be reproduced in any form or by any electronic or mechanical means including information storage and retrieval systems, without permission in writing from the author. The only exception is by a reviewer, who may quote short excerpts in a review.

Designed by Lauren Thorsen

Content edited by Fawn Gwynallen

Stansberry Research
Visit our website at www.StansberryResearch.com
Printed in the United States of America
First Printing: March 2018
Stansberry Research
ISBN- 978-0-9978333-5-5

About Stansberry Research

Founded in 1999 and based out of Baltimore, Maryland, Stansberry Research is the largest independent source of financial insight in the world. It delivers unbiased investment advice to self-directed investors seeking an edge in a wide variety of sectors and market conditions.

Stansberry Research has nearly two dozen analysts and researchers – including former hedge-fund managers and buy-side financial experts. They produce a steady stream of timely research on value investing, income generation, resources, biotech, financials, short-selling, macroeconomic analysis, options trading, and more.

The company's unrelenting and uncompromised insight has made it one of the most respected and sought-after research organizations in the financial sector. It has nearly 200 employees operating in several offices in the U.S. plus one in Asia, and it serves 350,000 customers in more than 120 countries.

Table of Contents

Foreword: This Controversial Market Is Set to Boom in 2018
By Nick Giambruno i

North America's Fastest Growth Sector
By Michael Ford 1

PART I

Conversations With the Experts
What Is This Marijuana Trend?

A Viable Supplement for Living a Healthy Life
Will Kleidon 19

Pioneering the Cannabis Business in America Today
Michael Lang 33

The Future of Legal Marijuana
Matt McCall 47

The No. 1 Factor for Success in the Cannabis Industry
Doug Esposito 63

This Man Set the Foundation for the Cannabis Economy
Aaron Salz 81

Canada's Most Connected Venture Capitalist
Matt Shalhoub 89

The CEO of the World's Largest Legal Cannabis Company
Bruce Linton 97

The Hidden Value in Legal Marijuana
Todd Harrison   105

PART II
Making a Fortune in Legal Marijuana
How the Millionaires Did It... and How You Can, Too

Marijuana Millionaires: How Regular Investors Are Cashing in on the Next Big Boom   125

The Potential Winners of the Global Marijuana Boom   141

Get Your Degree in Smart Trading Basics: Three Ways to Manage Risk   157

Putting It All Together   167

– Foreword –

This Controversial Market Is Set to Boom in 2018

By Nick Giambruno, Senior Editor, Casey Research

The U.S. is about to experience one of its biggest social and economic changes in generations.

And the story of Desert Hot Springs, California is the key to understanding it all…

Like many U.S. cities, Desert Hot Springs had chronic financial problems. The city filed for Chapter 9 municipal bankruptcy in 2001. Then it nearly went bankrupt again in 2014, when it declared a fiscal emergency.

At this point, the city's people were desperate to solve their intractable financial troubles. So they opened their minds to a new idea.

They voted to become the first place in California to allow indoor cannabis cultivation on an industrial scale. A full 70% of votes were in favor.

The city would, of course, tax these operations.

Local politicians had nixed the idea before. But the risk of two bankruptcies in 15 years changed their minds. Necessity has a way of doing that.

As we begin 2018, this once-dying city of 29,000 people – the forgotten neighbor of storied Palm Springs – is experiencing a renaissance.

Since the vote, it has issued permits to more than 30 growers for more than 3 million square feet of cannabis cultivation.

Hundreds of new jobs have been created. Previously vacant industrial real estate is buzzing with activity.

The local government now has a $15 million budget to pay its employees and run the city. Yet it's expected to take in $50 million in cannabis taxes alone.

Desert Hot Springs has gone from a dying pit stop with big money troubles to a booming city with more cash than it knows what to do with.

This is why widespread cannabis legalization is both inevitable and imminent: money.

Many places in the U.S. have similar stories.

Just look at Colorado. In 2017, the state's marijuana industry generated $1.3 billion in sales and $200 million in tax revenue. A decade ago, Colorado received zero dollars in marijuana taxes.

The industry also created 18,000 new full-time jobs. And that's just one state.

According to Bloomberg, the state of Washington will generate $155 million in cannabis-related taxes in 2018. After four years, that number is expected to exceed $1 billion.

In California – which started selling recreational cannabis legally on January 1, 2018 – cannabis taxes are projected to bring in at least $1.4 billion each year and create more than 100,000 new jobs by 2020.

In many of the states that have or will legalize cannabis, the tax revenue will exceed that of alcohol and tobacco. That's not something a cash-strapped state can turn its back on.

All that money will quickly become a permanent part of many state budgets. This will encourage other states to follow suit.

It's important to clarify something here...

The U.S. has a population of 320 million. Yet it has more people in prison than China, which has a population of 1.4 billion.

The U.S. also has a higher percentage of its population in prison than North Korea, Cuba, Iran, Syria, or any of the other countries it routinely demonizes.

The war on (some) drugs – and cannabis Prohibition in particular – is the biggest reason why the U.S. is still No. 1 at something: locking people up.

This pointless war has killed an untold number of people. It's put millions of nonviolent offenders behind bars. It's even ravaged entire countries, like Mexico and Colombia.

I'm not celebrating a new way for state and municipal governments to steal money from people voluntarily exchanging goods and services. I think taxation is morally indistinguishable from theft.

But taxing cannabis – rather than prohibiting it outright – means more personal freedom and less government coercion. It's a big step in the right direction.

Leaders around the world are rethinking marijuana Prohibition. Many longtime "drug warriors" are even changing their tune.

Prohibition is an unsustainable distortion in the cannabis market – a distortion that you can take advantage of to make massive profits.

Widespread Legalization Is Imminent

According to a Gallup poll, 64% of Americans say cannabis should be totally legal. That's the highest level of support since polling on this issue started 48 years ago.

In 2018, opposition to legalization has dropped to an all-time low of just 34%.

Meanwhile, support has more than doubled since 2000. And as you can see in the following chart, it's continuing to grow...

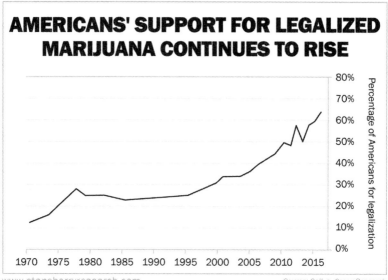

As of early January 2018, 29 states (plus Washington, D.C.) have approved legalized medical marijuana. And eight states (plus D.C.) have approved recreational use.

It's only a matter of time before other states do the same, especially those with budget problems. And eventually, the U.S. federal government will as well.

The economic benefits of legalization are simply too good to pass up.

It's absurd, but the U.S. federal government still classifies marijuana as a Schedule I drug. In other words, the feds put marijuana in the same category as heroin and LSD – and they consider it more dangerous than cocaine or meth.

But this ridiculous policy probably won't last much longer.

Of course, you might be worried that the Trump administration will shut down the legal cannabis industry – along with your chance to profit from it.

And there's good reason to be cautious. After all, Trump's attorney general did call marijuana "slightly less awful" than heroin.

Comments like this have stopped many people from investing in marijuana. But consider the bigger political picture.

First, the industry is too big to shut down. There's simply too much money and too many jobs at stake. The backlash would be overwhelming.

For example, now that recreational pot is legal in California, the U.S. legal marijuana market is expected to grow from $6.5 billion to $50 billion by 2026. *That would make it the same size as the American craft beer and chocolate markets combined.*

Second, states' rights matter to Trump voters. Cracking down on the industry would alienate the people who put him in office.

The new jobs and desperately needed tax revenue will make it politically impossible to roll back legalization.

Attacking the U.S. marijuana industry would cost Trump a lot more than it's politically worth. It wouldn't make any sense to do it.

The Birth of a $150 Billion Industry

The United Nations estimates the global cannabis market to be worth around $150 billion annually. That's very conservative.

For perspective, about $33 billion worth of coffee is produced annually. So the marijuana market is almost five times bigger than the coffee market.

It's also bigger than the iron, copper, aluminum, silver, corn, and wheat markets. You can see how these commodity markets stack up in the chart on the following page...

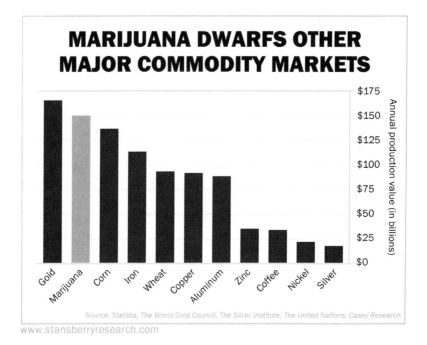

Until now, almost all of this money has been underground. But that's about to change.

The end of marijuana Prohibition means we can finally profit off marijuana without risking jail time.

Widespread marijuana legalization is inevitable. It's happening. And it's unleashing a $150 billion market that was once underground.

Those profits are up for grabs.

In the coming months, many investors will make life-changing fortunes as the marijuana market steps into the light.

For now, though, marijuana is still illegal in most places.

Up until recently, that made it virtually impossible for regular investors to cash in on the lucrative marijuana trade.

Prohibition has funneled billions of dollars in profits to drug lords, corrupt government officials, and thugs.

Those days are numbered...

The end of marijuana Prohibition in the U.S. became inevitable the moment California voters decided to legalize recreational use. Reversing this trend now would be more difficult than pushing a 20,000-pound boulder back up a mountain with your bare hands.

I've never seen an opportunity with as much profit potential as legal marijuana has right now...

Investing in the nascent cannabis industry right now is like investing in the beer industry at the tail end of Prohibition: Fortunes are going to be made. And you can be a part of that.

In this book, Michael Ford and the Stansberry Research team explore the cannabis industry – including why it's here to stay, what you can expect from this "new" commodity, and exactly how you can safely profit.

The experts indicate this is no speculative bubble. Cannabis is in the early innings of a major, long-term boom. Investors who get in on the ground floor could make an absolute killing in the U.S. "green rush."

Regards,

Nick Giambruno
January 3, 2018

– Introduction –

North America's Fastest Growth Sector

By Michael Ford

I still remember when the first marijuana dispensary opened near my house in Denver, Colorado.

Seemingly overnight, Santa Fe Avenue transformed from used car lots into a row of vibrant buildings, all advertising: "Medical Marijuana." Billboards proudly shouted advertisements for high-efficiency grow lights and hydroponic grow kits.

2012 was an exciting year in Colorado. What we didn't realize at the time is that we were about to lead the country in a widespread revolution that would change the way much of the country viewed a once taboo and misunderstood product.

Hi, I'm Michael Ford.

I come from a family of entrepreneurs. My father co-founded a multibillion-dollar business when I was a child, developing an entire industry in the days before the Internet. Growing up in that kind of environment – and now as a writer for Stansberry Research, one of the premier investment advisory publishers in the financial industry – business has always fascinated me.

And it's that mentality that drew me to explore the legal marijuana industry.

You see, love it or hate it... support the legalization and decriminalization of marijuana or not... the cannabis trend sweeping the globe these days isn't about hipster millennials throwing caution to the winds and taking over the world.

This is a legitimate, profitable business with incredible potential for both entrepreneurs and investors... as well as the countries they live in.

The legal marijuana market is one of the few opportunities in a generation where you can get in early on the next household-name industry.

This is like the invention of the computer... the birth of the Internet... and even the resurgence of the alcohol industry after Prohibition.

Business Insider reports that in 2016, legal marijuana sales in North America totaled just $6.7 billion, while the illicit market brought in more than $46 billion.

In other words, **more than 85% of marijuana consumers haven't moved over to the "legal" market yet**.

And there's still time to invest ahead of this massive consumer shift.

ArcView Market Research estimates that the legal cannabis business is on track to reach $24.5 billion in sales by 2021 – a 265% increase from early 2018 levels.

It also estimates that by 2025, the legal market alone will surpass $50 billion – a **650% increase** in less than a decade.

To put that in perspective...

Chicken and eggs are a $40 billion market.

Wine is a $38 billion market.

Dairy is a $35 billion market.

We're talking about a brand-new market that will soon outsell poultry, wine, and milk. And we're talking about a rare situation where regulations prohibiting marijuana sales are *evaporating* faster than merchants can open up shop.

Today, no single company has more than 1% of the market.

In other words, the big money is still up for grabs.

There are millions, if not billions, of dollars to be made from the legalization of marijuana in the United States alone.

That may explain why, in 2017, the top companies in this space saw their share prices soar by nearly 300%. Some individual names climbed 1,000% or more.

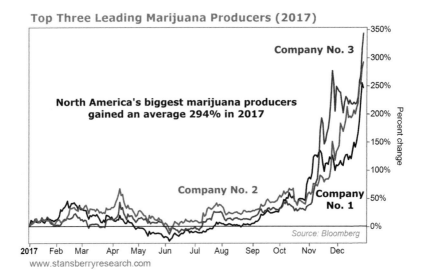

But you haven't missed out...

This isn't a fad or a quick bubble. Despite opposition from the highest levels of government, industry insiders say the legal marijuana industry is here to stay.

Speculating on marijuana stocks isn't right for everyone. And we do not actively recommend any of the companies analyzed in this book. But if you decide this industry is right for you, it could be incredibly lucrative for disciplined investors...

Take Terra Tech, for example. It has interests in 12 subsidiaries in the marijuana industry, including businesses focused on grow-operation technology and cannabis-permitting technology.

I mention it because one of the company's subsidiaries, Blum – a cannabis dispensary in Oakland, California – may be the most profitable single store location in the world by revenue.

The dispensary churns out an astounding $7,000 per square foot.

That's 25% more income per square foot than the average Apple Store.

To put that in perspective, the Apple Store makes more money on a square-foot basis than any other publicly traded company in the world.

Terra Tech is not a company I'd tell you to run out and buy today. The Blum dispensary represents just a small part of Terra's portfolio.

But the revenue this single marijuana dispensary generates speaks volumes about how much stands to be made in the coming years as the industry matures.

Our job is not to guide your moral compass... It's to help you grow your wealth. Legal marijuana is one of the most dynamic and controversial markets in America. Many companies in this burgeoning industry have the potential to generate thousands of percent returns for early investors. But the sector is also rife with risk, misinformation, and hype.

In this book, we'll detail where you can find legitimate opportunities, what significant challenges still confront the industry, and which critical pitfalls to avoid.

So What Is the 'Legal Marijuana Industry'?

Each year, consumers around the globe spend an estimated $150 billion-$200 billion on their marijuana habit. Until recently, most of these sales took place on the "black market" because marijuana was illegal in nearly every country. But this hugely profitable industry is finally stepping into the light...

In 2000, marijuana was illegal in nearly every jurisdiction on the planet. In 2018, 29 countries allow some form of marijuana consumption.

A legalization movement is sweeping the globe, setting the stage for a massive new industry.

The movement started when countries like Portugal began experimenting with decriminalization of marijuana possession. (Drugs were still illegal, but the sentence for possession was reduced to a small fine instead of jail time and a criminal record.)

After decriminalizing drugs in 2001, overall drug use in Portugal steadily declined. The prison population shrank. Drug-related offenders accounted for just 21% of inmates in 2012 compared with 44% in 1999. And Portuguese drug-related deaths fell to the second lowest in the European Union.

Other countries like Canada began legalizing marijuana for medicinal use back in 2001. These early advocates paved the way for a global movement.

Even Germany – which has some of the strictest narcotics laws in Europe – legalized medicinal marijuana in 2017. This move from Europe's largest economy shows that no country is immune from the trend toward marijuana legalization.

Many expect Italy will follow in Germany's footsteps, which could spark a chain reaction throughout Europe.

Legalization is spreading across South America and the Asia-Pacific, too. Uruguay became the first country to fully legalize marijuana (medicinal and recreational) in 2013. Brazil decriminalized marijuana in 2006, followed by Mexico in 2009. Australia legalized medicinal marijuana in 2016. New Zealand will hold a national referendum on full marijuana legalization by 2020.

In the U.S., Colorado and Washington became the first states to legalize recreational marijuana in 2012. By early 2018, more than half of U.S. states allowed medicinal or recreational marijuana consumption.

Despite the groundswell of support among individual U.S. states, marijuana remains illegal on the federal level. *This means that federal agents can arrest users and producers of marijuana, even if they're operating legally under their state laws.* **All indications suggest this will soon change**...

Gallup polling data shows the majority of Americans support marijuana legalization. This is a relatively recent phenomenon. In 1996, only 25% of Americans favored legalization. That number soared to 64% in 2017.

In other words, support for marijuana legalization isn't just growing... its growth is accelerating.

Despite the shift in global attitudes toward legalizing marijuana, skeptics abound...

The naysayers compare today's enthusiasm in the marijuana industry with the cryptocurrency mania. But unlike cryptocurrencies, marijuana isn't based on a speculative rise in some intangible asset. The companies in this space don't need to build demand from scratch. The multibillion-dollar global marijuana industry *already exists* – it's just moving from the underground "black" market into a legal "white" market.

We haven't seen such a massive industry emerge from the black market since the end of alcohol Prohibition in the 1930s. The indisputable failures of alcohol Prohibition show why marijuana legalization is all but unavoidable.

Alcohol Prohibition: A Failed Idea

In 1920, the U.S. government passed the 18th Amendment, which outlawed the production and sale of alcohol. Proponents of Prohibition argued that making alcohol illegal would...

- Solve social problems
- Reduce crime rates
- Decrease the prison population
- Lower taxes by reducing the need for police and prison systems

Just 13 years later, the outcome was clear: **Alcohol Prohibition was one of the biggest national policy disasters in American history**. The government admitted defeat and ended Prohibition in 1933.

On paper, Prohibition sounded like a good idea. But in practice, the policy created more problems than it solved.

You see, governments cannot eliminate demand. Making high-demanded products illegal doesn't make them disappear... It only creates a lucrative opportunity for criminals to supply the market.

In the 1920s, alcohol was America's fifth-largest industry. People weren't going to suddenly stop drinking. So all Prohibition did was effectively hand this massive economic profit center over to criminals, fueling the rise of organized crime in the country. Businessmen in suits and ties were replaced with gangsters like Al Capone and Machine Gun Kelly.

The emergence of large-scale criminal enterprise sparked a 78% rise in America's homicide rate during Prohibition. More crime meant more arrests. The federal prison population more than quadrupled. By 1930, half of the new prisoners each year were charged with violating Prohibition laws.

Well-financed criminal organizations overwhelmed local police forces. Entire new federal agencies were created to deal with the surge in criminal activity. And since bootleggers and gangsters don't pay taxes, the government missed out on valuable tax revenues that would have otherwise come from the legal liquor industry.

So much for solving social problems, reducing crime rates, decreasing the prison population, and lowering taxes...

Instead of improving the health of alcohol consumers, Prohibition created new threats to health and safety. Criminal organizations don't abide by the same rigorous health and safety regulations as legal businesses. So liquor poisoning deaths soared four-fold during the first few years of Prohibition – from 1,064 per year in 1920 to 4,154 per year in 1925.

History is clear: Alcohol Prohibition actually boosted crime, exploded the prison population, and made alcohol more dangerous. Plus, it diverted economic profits and tax revenues into the hands of violent criminal organizations.

American satirist H.L. Mencken best described the failure of alcohol Prohibition in 1925 (only five years into the program)...

> Prohibition has not only failed in its promises, but actually created additional serious and disturbing social problems throughout society. There is not less drunkenness in the Republic, but more. There is not less crime, but more... The cost of government is not smaller, but vastly greater.

Despite this real-world case study in the failures of alcohol Prohibition, the U.S. and most other countries pursued the same flawed policy with marijuana in the 20th century.

Marijuana Prohibition: Repeating the Mistakes of History

Just four years after the end of alcohol Prohibition, the U.S. government came down with a severe case of amnesia...

In 1937, Congress passed the "Marihuana Tax Act of 1937." (That's not a typo. The arcane "marihuana" spelling started getting phased out of popular use in the 1960s, replaced by the modern "marijuana.")

This act effectively outlawed the drug. It was repealed in 1969. But marijuana became illegal again by 1970 when Congress passed the Controlled Substances Act of 1970 as part of President Richard Nixon's broader "War on Drugs." That act labeled marijuana as a Schedule I drug... the most dangerous classification, with "no medical value."

Like alcohol Prohibition in the 1920s, making marijuana illegal created a new class of criminals. After remaining stable for decades leading up to the early 1970s, the U.S. prison population has surged more than 400% since the War on Drugs began. This massively outpaced the general population growth of about 50% during the same period.

As of 2014, the United States housed the largest prison population on the planet – ranking five to 10 times higher than other democracies around the world. We can point directly to marijuana Prohibition as a leading cause. *Each year, police arrest more people for marijuana possession than for all violent crimes combined.*

This crackdown on marijuana costs taxpayers up to $20 billion each year, according to one Harvard economist. Unfortunately, we have nothing to show for this massive allocation of resources...

With decades of data under our belts, studies show that no correlation exists between drug arrests and drug use. In other words, **putting people in jail for consuming marijuana does nothing to slow the rate of consumption.**

Marijuana consumption has soared in the last nearly five decades. A 2017 Gallup poll shows that 45% of Americans report having tried marijuana in their lives. That number was only 12% in 1973.

Even if you disagree with marijuana legalization, it's clear that the drug does not belong in the Schedule I category. Unlike other Schedule I drugs, like heroin and LSD, it's practically impossible to overdose on marijuana... even according to those in charge at the U.S. Drug Enforcement Administration (DEA).

In a 1998 court case, a DEA judge argued that marijuana should be removed from the Schedule I classification. The reason? You would need to consume the equivalent amount of THC contained in approximately 20,000 to 40,000 joints (marijuana cigarettes) to overdose. In the words of Judge Francis Young...

> A smoker would theoretically have to consume nearly 1,500 pounds of marijuana within about 15 minutes to induce a lethal response.

Despite humans consuming the drug for millennia, we have yet to record a single death from marijuana overdose. Yet somehow, marijuana is legally classified as more dangerous than Schedule II drugs like cocaine and methamphetamine.

There's only one group who benefits from making marijuana illegal: Criminals. Just like with alcohol Prohibition, illegal marijuana funnels massive sums of money into the world's most violent criminal organizations. I'm talking about Mexican drug cartels who, until recently, supplied the majority of marijuana to U.S. consumers.

All efforts to combat Mexican drug cartels with force have failed. Beginning in 2006, Mexico's then-president Felipe Calderón mobilized the country's military to rein in the cartels. But this only amplified the violence. Since then, Mexico's drug war has produced more than 200,000 casualties. Plus, an estimated 30,000 missing persons.

If you legalize marijuana, you destroy the cartels and the crime that comes with them.

The Benefits of Legalizing Marijuana

The paradox of Prohibition is rooted in basic economics.

The more drug dealers you take off the streets, the more supply comes off the market. This sends prices up, creating a greater incentive for other criminals to come in and fill the vacuum in supply. Like the mythical Greek water serpent Hydra, each head you remove in the illegal drug trade only produces two more in its place.

The proven way to eradicate criminal organizations is to attack their profit centers. We've seen early signs of this in the U.S., as a flood of new supply has hit the market from state-legalized producers.

From 2008 to 2015, legal U.S. growers took away half of the Mexican market share and pushed prices down. A December 2014 NPR interview with a Mexican marijuana grower highlighted how American decriminalization efforts could put him out of business...

> Two or three years ago, a kilogram of marijuana was worth $60 to $90. But now, they're paying us $30 to $40 a kilo. It's a big difference. If the U.S. continues to legalize pot, they'll run us into the ground.

Not only will legalization take money away from criminals, but it will also generate economic growth and billions in tax revenues. Studies show that U.S. legalized marijuana could generate up to $132 billion in tax revenues and create more than a million jobs by 2025.

You can't find a safer play than betting on politicians exploiting new sources of tax revenue. The promise of billions in marijuana taxes all but ensures the industry eventually goes legal around the globe.

Even many Republican lawmakers have reversed their stance against legalization. The clearest example of this came on the heels of an announcement made by U.S. Attorney General Jeff Sessions.

On January 4, 2018, Sessions rescinded an Obama-era Justice Department guideline known as the Cole Memo. This memo set a precedent of noninterference from the federal level into state-sponsored marijuana legalization, despite marijuana's illegal status under U.S. federal law.

In other words, the Cole Memo protected marijuana producers who operated within the legal constructs of their home states. Removing these protections sparked a backlash from voters and political leaders across both sides of the aisle. Fifty-four members of Congress signed a letter asking Sessions to respect the will of voters and back off his anti-marijuana efforts.

One Republican congressman from Colorado – Mike Coffman – publicly questioned whether Sessions understood the U.S. Constitution. Less than 10 years earlier, Coffman staunchly opposed marijuana legalization. He changed his position and became an ardent supporter of legalization after Colorado legalized marijuana in 2012.

The growing shift among lawmakers and citizens shows no sign of stopping.

If the U.S. federal government truly reflects the will of the people, it won't be long before marijuana is legalized in America. If not, we will surrender an enormous economic opportunity to more progressive and socially aware countries, like Canada.

Canada: Ground Zero for Cannabis

Canada took the lead in the global marijuana market in 2001, when it became the first country in the world to legalize medical marijuana. In 2018, Canada has the most mature legal cannabis industry on the planet.

When running for Prime Minister, Justin Trudeau underscored the philosophy behind Canada's push toward legalization...

> Criminal gangs and street gangs are making millions of dollars of profits off the sale of marijuana, and we need to put an end to this policing that does not work.

When elected in 2015, Trudeau followed through on his campaign promises to move Canada from just medicinal legalization to full-scale recreational legalization. Previously, only Canadians with qualifying medical conditions could buy marijuana for medicinal use. But the plan is that starting on July 1, 2018, marijuana will be legal for any adult over the age of 21 to consume recreationally, without a doctor's prescription.

This will unleash a massive new wave of demand in Canada.

With recreational use, experts project the Canadian cannabis industry will grow to about C$8 billion-C$10 billion in size... but this is only the start. The real prize lies in the $150 billion global market via export opportunities. For comparison, the total value of Canadian oil exports is around $50 billion.

Canadian companies have a unique opportunity to capitalize on the massive global market. You see, despite the global push toward legalized marijuana consumption, few countries have established a robust framework for marijuana production and exports. In 2017, only two of the more than 20 countries with legalized marijuana even allowed for its export – Canada and the Netherlands.

Because Canadian lawmakers have been regulating the industry since 2001, Canadian companies have a relatively clear and effective system to operate within. This has allowed Canada to become the world's No. 1 marijuana exporter. The country has a huge lead...

Many countries can't even supply their own domestic markets. Consider Germany... With a population of 83 million people, experts expect Germany will become a leading demand center for medical cannabis. But due to strict regulations on cannabis production, Germans rely on imports to satisfy their domestic demand. Canadian producers have stepped in to fulfill this demand, becoming the leading providers into the German market.

Germany is just the start. As of early 2018, seven Canadian companies have been granted export licenses to sell Canadian marijuana to a growing list of countries around the world, like Australia, New Zealand, and the Czech Republic.

Canada is emerging as the global leader in the burgeoning marijuana industry.

How to Play the Coming Global Pot Boom

As with any new trend, there will be many people who go about it the wrong way and lose money. That's why the profit potential is so high for the folks who understand how this market really works.

In *The Marijuana Industry Insider's Playbook*, we speak with some of the leading entrepreneurs in the legal marijuana space to learn about the basics of this industry, the longevity of the cannabis boom, and where to consider investing.

If you want to gamble… this book isn't for you.

If you are morally opposed to marijuana… this book isn't for you.

But if you want to make money in a promising new industry that is only in the first innings of what could be a massive migration of wealth… you need the insider's advantage.

In these pages, my team and I have assembled the insights, tips, and strategies you'll need to get started – right out of the mouths of the businessmen and women who are leading the charge.

As with any new industry, you can find almost any "angle" to legally make money, as long as you know the secret to doing it correctly…

The successful people I encountered on this journey didn't wait for opportunity to find them. And they didn't invest for the present. Each of them looked to the future of where the industry was going.

That's what makes the rise of a new marijuana industry so intriguing. Though this trend has already minted its fair share of millionaires, we're still in the early days. And if you know what to look for, what to avoid, and how to understand these businesses... you may enjoy huge profits, too.

In the following chapters, you'll hear from eight entrepreneurs in the legal marijuana industry. They each share their unique take on the state of this market today, as well as what we can expect in the future... and why this boom is here to stay, despite government opposition.

We'll also introduce you to several ordinary folks who have already made a fortune in the legal marijuana industry. We'll show you the top stocks to watch as this trend unfolds. And we'll share some basic investing tips you must follow to succeed in the markets.

I hope when you've finished this book, you are as excited about this opportunity as I am.

PART I

Conversations With the Experts
What Is This Marijuana Trend?

– Chapter 1 –

A Viable Supplement for Living a Healthy Life

A conversation with Will Kleidon
Founder and CEO, Ojai Energetics

December 27, 2017

Will Kleidon founded Ojai Energetics in 2014. His goal was to provide the purest, healthiest, and most ethically produced cannabidiol (CBD) product on the market... giving people a viable supplement for living the most effective and healthiest life possible.

Four years later, the small business is a leader in cutting-edge, science-driven organic nutraceutical and medical food products for achieving optimal physical and spiritual health.

Will has collected one of the most interesting executive boards in the world, including famous athletes, doctors, and well-connected members of the entertainment industry.

Prior to founding Ojai (pronounced "oh-hi"), Will coordinated a sustainability festival that attracted 25 organizations and 500 attendees. He went on to study permaculture in Australia, where he apprenticed with one of the world's leading experts, Professor Robyn Francis, at Permaculture College Australia. There, Will earned the second-highest nationally accredited certification in Permaculture Design.

In the following interview, Will discusses...

- The components of the cannabis plant, including the difference between hemp and marijuana.

- Why legal marijuana is going to be one of the biggest-growing economies and industries globally in the next decade.

- The biggest opportunities for marijuana to disrupt the markets.

- And the banking problem cannabis companies like his face in today's political climate.

Stansberry Research: Will, thanks for taking the time to meet with me. I look forward to hearing about your company and your thoughts on the industry in general.

You are the founder and CEO of Ojai Energetics, based out of California. Can you tell me a little bit about when and why you started the company?

Will Kleidon: Sure. I was born and raised in Silicon Valley. We moved all over and ended up in Ojai, Southern California. I've been there ever since. I decided to start the company officially at the end of 2013/beginning of 2014 out of a personal need. I was looking for CBD oil for general anxiety around that time.

Stansberry Research: Can you explain what CBD oil is for our readers who may not be as familiar?

Kleidon: Of course. To put it simply, the cannabis plant is made of a variety of chemical compounds called cannabinoids. There are more than 100 cannabinoids in the plant. There's evidence that each can be used for relieving symptoms of various illnesses. But most people only hear about one: THC.

Stansberry Research: That's the psychoactive compound in the cannabis plant.

Kleidon: Correct. Even if your readers don't use marijuana, they've probably heard of THC. That's what gives you a "high." In medical marijuana, THC has been used as an anti-epileptic, anti-inflammatory, antidepressant, and lowers blood pressure.

Cannabidiol (or "CBD") is a cannabinoid like THC, except it is non-psychoactive and has a variety of health benefits. Research shows that CBD is an anti-inflammatory, anticonvulsant, antioxidant, and antipsychotic agent.

Stansberry Research: Thank you for explaining that. So back to your story... You started Ojai because you were looking for CBD as a treatment for anxiety...

Kleidon: Yes. I thought I needed to get it through a dispensary. But when I looked online for dispensaries that sold CBD, I was shocked to see that I could purchase it directly through Amazon. I ordered it, and was even more shocked when it showed up in the mail.

But when I looked at the ingredients, the CBD was filled with crap and fillers and synthetic ingredients. I did some due diligence on how this could be legal. At that point, it was pre-Farm-Bill. So the CBD oil had to be from stocks and stems only of hemp grown outside of the U.S. That has since changed.

[**Editor's note**: The 2014 Farm Bill allows the growth of industrial hemp for authorized research purposes if state law permits.]

But it was at that point that I knew I could create a superior product.

Back then, there was a lot of concern about where companies were sourcing their hemp from. Primarily because a lot of it was sourced from China and had a high metal content. I wanted to start something organic and clean... something I felt good taking and giving to friends and family. This was especially important with the trend of people wanting more transparency in food and other products they consume.

That was kind of the impetus for my company. I got the green light from my legal team and just took off from there.

Stansberry Research: Your business is obviously derived from hemp plants... how is hemp any different than marijuana?

Kleidon: There's a lot of confusing information out there. And at first, I was even confused about what distinguished hemp from cannabis.

The agreed-upon definition for hemp is: "Any cannabis plant whose THC is below 0.3% on a dry weight basis." That definition is more or less standard throughout Europe and the rest of the world.

It's a legal distinction, not really a botanical distinction. A good way to think of it is that just like there's different varieties of tomatoes, there's different varieties of the cannabis plant.

The hemp definition is meant, functionally, to designate between psychoactive and non-psychoactive cannabis.

So when we talk about hemp, we are talking about a plant that has historically been used for food and fiber. When we talk about marijuana, we are generally talking about the high-THC cannabis plants, which are commonly used for medicinal psychoactive purposes.

Stansberry Research: A lot of investors think this legal marijuana boom is just the current investing fad. In your own words, why do you think the industry's here to stay?

Kleidon: Legal marijuana is going to be one of the biggest-growing economies and industries globally in the next decade.

First of all, humans have been cultivating, working with, and consuming the cannabis plant since the beginning of agricultural society. Polynesians brought it wherever they traveled. The Romans brought it everywhere they colonized.

It was actually required by law to grow hemp in Virginia during colonial times. The first draft of the Constitution was written on hemp.

At one point, Thomas Jefferson was going to patent an innovation of processing the hemp fibers to make it into cordage and paper. But he decided not to because he believed it was too important for society to be patented.

Cannabis, but specifically hemp, was at the center of economies for millennia until Prohibition banned the plant in America in the 1930s. But now that it's back in play, it could disrupt multiple industries.

Stansberry Research: Where do you think the biggest opportunities for disruption are today?

Kleidon: The psychoactive element of THC makes it an attractive alternative to alcohol for recreational use. There are no calories, which makes it very attractive to people who are dieting or who are health conscious. But it's also very predictable, so people would have more control than with alcohol consumption. There's no such thing as a toxic overdose of marijuana.

So that's the biggest market most likely. It will probably outgrow alcohol sales. And there's already evidence that indicates alcohol sales are lower in areas that allow recreational cannabis sales. Not to mention, this will receive a giant tailwind as the taboo surrounding cannabis consumption continues to go away.

As for CBD and the other cannabinoids, you are going to see huge disruptions in health and wellness. Not everyone wants to get high, but everyone wants to live a longer, optimized, and healthy life.

What most people don't know is that CBD is a vital micronutrient that people have been consuming for millennia. It feeds the endocannabinoid system, which runs every other system of the body. And I believe every single human needs daily dietary intake of CBD to have a healthy endocannabinoid system.

CBD isolated by itself is fairly limited in the scope of its benefits compared with CBD that works in combination with the hundreds of other compounds the plant produces. Our ancestors didn't consume isolated CBD. They consumed CBD with the full spectrum of compounds. So when I say "CBD," I'm speaking of CBD with a broad spectrum of other compounds... not "CBD isolate."

The actual implications of having a healthy endocannabinoid system are profound in terms of regenerating brain cells and overall homeostatic balance, which means making sure that every single system in the body is optimized. So on the health and wellness side, it's probably one of the most disruptive pieces for modern humans for optimized health.

On the medical side, the story is pretty much the same. Because the endocannabinoid system runs every other system of the body, the amount of benefits for multitude of issues are profound. The science behind some applications, like we've seen with Dravet syndrome – or severe myoclonic epilepsy of infancy – are amazing. To see a child with severe epilepsy go back to living a normal life because of a compound in the cannabis plant is fascinating stuff.

Stansberry Research: This is a commodity that we've used and traded and relied on for thousands of years, despite the Prohibition of the last 80 years or so. Is it safe to say we're still in the first or second inning of this kind of movement?

Kleidon: Yes. We're just scratching the surface of what we can do with this plant and how it can disrupt all kinds of established businesses. There's beginning to become a maturity in terms of people who are doing it properly and are beginning to rise to the surface. But it really is the beginning of a catalyst for, in my opinion, probably the most innovative and positively disrupting spaces on the planet. So we're just at the beginning.

Stansberry Research: What does it feel like to be involved in an industry that's so new?

Kleidon: It's amazing. There's frustrations, and with that come major opportunities and challenges. And it's absolutely exhilarating and rewarding. Every day is exciting. Because not only are we on the cutting edge of health, but we're in a completely emerging market. It's a lot of fun.

We get to change lives and simultaneously set the standard for an emerging market. By doing it right, by working with the right people, we can lead the market. That's rewarding.

Stansberry Research: As a small business owner, do you ever worry that some positive federal resolution of the marijuana debate will allow big multinationals to come in and squeeze out some of the smaller players?

Kleidon: There is a lot of concern that bigger players will come in and dominate the space. But that doesn't mean smaller players will disappear...

You'll always have people who will go to a big store and get their generic boxed wine or their big-agriculture, cookie-cutter cannabis. But there's also going to be an exponentially growing demand for boutique cannabis. That's especially true with the millennials driving the market by 2020. They want stories. They want transparency. They want clean practice.

So yes, there will be major business coming in. But it is really only a concern if you have no competitive advantage or intellectual property (IP).

For example, what makes my company unique is that instead of just going to market immediately, I partnered up with law firm Wilson Sonsini Goodrich & Rosati ("WSGR") in 2014. WSGR was recently ranked No. 1 for biotech and IP firms globally.

The first thing I did was secure a patent catalog. It's for all cannabinoids and different delivery systems, disease claims, industrial applications, and more. One thing that we pioneered at Ojai was figuring out how to make the oil completely soluble in water by only using certified organic ingredients. I wanted to corner all the abilities of doing that with cannabinoids without using synthetics. And that's just a sliver of what we've built out with this catalog.

Because the quality of our IP is so strong and because of who our IP attorneys are, the two biggest alcohol companies just approached us and said, "We're ready. We're super interested. We want to start experimenting."

Stansberry Research: Wow, so you see applications of CBD in drinks going forward? Are beverages going to be a big area of the market?

Kleidon: Yes, 100%. Because of our solubility and how soon after consumption our product begins to take effect, we've got some of the biggest players approaching us right now...

We're working with one of the leading coffee companies. CBD actually gets rid of the jitters you get from drinking coffee and protects the body from caffeine over-targeting the adenosine receptors, so it actually makes a safer and much more enjoyable cup of coffee.

The biggest kombucha company selected us. And we have major distributor pipelines and big boxes that are ready to jump – 7-Eleven, Kroger, and Costco are all gearing up. On the alcohol front, it makes a safer drink too, so you can imagine how big that market could be by itself.

And don't forget the many benefits on the health side...

Biochemically, having a healthy endocannabinoid system puts you into a flow state – like the focus athletes feel when they're focused and "in the zone" – and optimizes your executive functioning. For example, we've got a whole team of professional athletes we work with. We just met with dieticians from the Los Angeles Rams football team and the Los Angeles Dodgers baseball team. That tells me it's rapidly going mainstream.

Stansberry Research: You see applications in everything from soft drinks to coffee to beer to liquor to overall wellness.

Kleidon: Yes. But there are so many applications I haven't even mentioned. For example, we can turn our waste products into concrete replacements that are bulletproof, fireproof, and waterproof.

Stansberry Research: You mean you can use the plant cellulose byproduct from your production process?

Kleidon: Right. We're working with the city of Lancaster to demonstrate different building materials. We can even 3D-print houses out of it.

Stansberry Research: Do you anticipate anything in the next couple years that you think will serve as a major catalyst for growth or change in the industry?

Kleidon: In 2018, it's a little bit of a cluster with the legal framework in the U.S. I think we're going to see federal legalization coming down the pipeline for non-hemp cannabis.

Stansberry Research: On the medical side? Or on recreational, too?

Kleidon: On both. There's just too much money in it to ignore. Look at the success of Colorado. And with California coming online, I think the sixth-largest economy on the planet, it's simply a huge boon for anyone even tangentially involved. And don't forget that Canada's jumped on board. You've got other countries where medical and recreational marijuana are federally legal. So I think we're just at the tip of the exponential curve.

Stansberry Research: If you had to give me a ballpark, what timeline do you think we're looking at for federal legalization?

Kleidon: I'd say within the next five years.

Another big catalyst that's more near-term that folks should keep in mind is when the numbers start coming out of California after the first quarter of 2018... Medical marijuana has been legal in the state for years. And as of January 2018, recreational marijuana is now legal as well. The amount of economic growth that will be reported in the headlines is going to be impossible to ignore. And you can imagine that investments in the space will soar accordingly.

Stansberry Research: Is there anything coming down the pipeline specifically for CBD that is going to impact your business?

Kleidon: Yes. So, right now there's confusion around the FDA's opinion on whether you can sell CBD as a food or a supplement – and actually call it CBD. We expect it is going to get clarified very soon. Right now, it's not focusing on CBD. It's just citing a "full-spectrum cannabinoid-rich hemp extract."

Walmart's selling it only as "full spectrum." Amazon will sell it as long as you don't call it CBD. We're about to roll out little energy shots and health shots into 7-11. CVS is ready to jump.

Stansberry Research: Any other events investors could potentially trade around?

Kleidon: Yes. Another big catalyst I expect in the near term is a resolution of the banking problem.

Stansberry Research: Can you tell us what you mean by "the banking problem"?

Kleidon: So in most states where marijuana is legal right now, it's very difficult for marijuana businesses to get access to normal banking services, like opening a checking account.

That's because banks in our country are insured by the FDIC, and the FDIC is obviously a federal organization. Because cannabis is still considered a Schedule I narcotic at the federal level, these banks risk their entire operations by dealing with businesses that deal with cannabis. So right now, most entrepreneurs deal almost entirely in cash or have to come up with creative banking solutions.

The one area where a lot of folks are going to put their money is local credit unions, but they can't keep up.

Stansberry Research: Do you encounter that despite the fact that you aren't touching or selling products with THC?

Kleidon: It's a nightmare, yeah. The issue has been with Visa and MasterCard. But there's a big lobby push. The banking solution will clear up.

Technically, there is no legal issue at a federal level if you're processing hemp CBD at all. As long as it is grounded in compliance with the Farm Bill. But banks are not necessarily the most creative in their interpretation of the rules nor are they risk takers. They essentially issued a blanket statement that they don't want to underwrite anything because if it isn't in compliance with the Farm Bill, the product is a controlled substance.

As a result, the European market is absolutely exploding. We're banking in Europe. It's a pain. I mean we grew the company by 60% in 2017, and it's arduous to get a checking account in the United States. But that's going to clear up too, which will make it a lot easier to do business.

Stansberry Research: Are you personally invested in any other areas of the market?

Kleidon: Not currently. There's one play I've got around credit card processing that I own through Ojai. We own 40% of it.

Stansberry Research: Is that a marijuana-oriented play?

Kleidon: It applies to any cannabis plant or derivative. The way I like to approach IP is to set it up so it applies to the entire cannabis plant, not just hemp. We're currently focused on hemp because it's federally legal, so that's all we'll do in the U.S. But when it becomes legal in the U.S., we want our existing patents to cover that future market, too.

And we'll be doing medical and recreational products in Canada and other countries through where marijuana is legal. We can do that by licensing our patents so that we're not violating any federal law, even though our company is here in the U.S.

Stansberry Research: So clearly, you think that's a big investible area going forward.

Kleidon: Yes. Processing is one area where there's huge opportunity. I'd say the safest ways to play it is to find companies that have intellectual property that's done by proper IP attorneys obviously. Because if it's not good, it's not good.

Compliance is going to be massive too. So seed-to-sale. That means making sure the marijuana is tracked and accounted for throughout the supply chain. These businesses aren't actually touching it, yet they're working with the industry.

If you're wanting to go into medical cannabis, I'd say that's a very safe investment – seed-to-sale processes that have contracts with the states. That's not going to go away. Neither will the market for innovative lighting solutions for indoor grows.

Stansberry Research: What do you look for specifically when you are looking at a company's IP?

Kleidon: I like to make sure that the company's built for the future, not just the present moment. That's the most critical part to get right.

One great example is people investing in CBD-isolate companies.

Stansberry Research: What's an isolate?

Kleidon: I mentioned this briefly earlier, but CBD isolate is the single-molecule CBD. It contains no THC. And it can be easily used at home to create your own edibles.

The plant produces over 100 cannabinoids, possibly more. And then there's polyphenols, phytonutrients, terpenes – we refer to it as a full spectrum that the plant produces. Again, that full spectrum is what we've evolved eating, through direct consumption or through eating animals that were fed it as animal fodder.

But when we isolate the CBD compound, it loses what's called "the entourage effect." It becomes limited in its bioavailability and its effect and action on the body.

So people are jumping into these CBD-isolate companies because the market's hot for it. What they're not seeing is that that CBD isolate does clearly violate the 201(ff)(3), which is the FDA's opinion. And it will most likely be classified for pharmaceutical use only.

Stansberry Research: Why do you think this corner of the market has gotten so hot? What are these isolates used for?

Kleidon: They're currently used for consumer-based products like foodstuffs. But trust me, it's not going to stay in our food. It's going to be a pharmaceutical-only application. But I see a ton of investors jumping in on these brands before that happens because they're hot and trending and folks are getting returns.

I expect in under a year that whole space will shut down for isolated CBD. The brands can adapt and do other things, of course. But if they're making CBD-infused water, they're going to have to license from us.

There are a few other companies that can do water-soluble full-spectrum, but they have to use a nano petroleum, which is just not healthy, and they'll never have the USDA stamp. That's one example of where people are jumping in but they're not seeing the direction the regulatory environment is heading in.

Stansberry Research: So that's something that needs to be navigated by somebody who knows what they're doing.

Kleidon: Exactly. Do your due diligence. Being in an emerging market, there are a lot of charlatans out there. I would say just do really thorough due diligence and make sure that you're looking at what their future play is and not just short-term return.

– Chapter 2 –

Pioneering the Cannabis Business in America Today

A conversation with Michael Lang
Founder, PharmKent LLC

February 13, 2018

Michael Lang is a seasoned executive and serial entrepreneur. In the late 1970s, Michael served as the manager for derivatives trading at investment bank Alex. Brown & Sons, where he pioneered the use of real-time computer systems to analyze stocks and commodities.

His career has followed a path of information science, including nearly a decade at financial news and data provider Bridge Information Systems, where he broke ground on the delivery of real-time analytical information for stocks and options.

Michael has also served as a senior vice president at international news agency Reuters, the president of industrial equipment manufacturer Symmetrical Technologies, and a member of the executive team at Vitria Technology. During his 40-year career, Michael has served on the board of seven major technology firms. Today, he's a managing director at Stansberry Research.

In the last 20 years, Michael has built and sold three companies to public enterprises.

In 2018, he is launching another new business called PharmKent LLC, this one in the cannabis industry... Michael is one of the few people licensed in Maryland to operate a medical marijuana dispensary, which he founded, capitalized, and operates with a group of invested partners.

Michael is an expert at navigating and complying with the Maryland medical marijuana regulatory framework and is proud to help bring a novel treatment to patients across Maryland.

In the following interview, Michael shares his first-hand experience starting a new cannabis business in America today. He discusses...

- The sort of obstacles new cannabis companies can face.

- The one thing that stops a lot of folks from diving into cannabis investments... and why those concerns are completely unfounded.

- Why the political risk posed by Attorney General Jeff Sessions and a Republican-controlled Congress is not a threat to legitimate cannabis businesses.

- Why the cannabis industry is insulated from a market decline.

- And a stream of opportunities that could be big winners over the next decade.

Stansberry Research: Michael, before coming to Stansberry, you were a veteran software executive. As far as I can tell, you've had a very successful business career, and you don't strike me as the kind of guy who has any interest in using cannabis recreationally. So can you share with our readers how your career led you along this path, what made you decide to move into this space, and why?

Michael Lang: Primarily, just a great business opportunity. Secondarily, I was a chemistry major in college. And actually, my first job, after I got out of college, was pharmaceutical sales rep for pharma manufacturing firm Upjohn. So I know an awful lot about medicines.

I do believe [cannabis] is a therapy that's useful for a range of indications. But primarily, this is a great business opportunity.

Stansberry Research: I know that your businesses are here in Maryland. Maryland allows medical cannabis use, but recreational use continues to be prohibited. Can you tell me a little bit about what the process looked like to get everything going?

Lang: Well, the challenges were significant. Maryland legislature passed a law in 2013 that created a commission to write the regulations for the medical cannabis industry in Maryland. The commission took nearly two years to write extensive regulations. Once those regulations were published in 2014, a procedure for applying was described.

So I went through the process, along with my partners, of filling out the application and supplying the massive amount of information about each individual partner that was required to even submit the application. We applied for a grower, processor, and dispensary license. We failed on the first two, but we were awarded a dispensary license on December 1, 2016.

Since then, we've been jumping through regulatory hoops to supply the additional information Maryland needs to get through Phase II application. Phase I application got us the award of a license. We needed to do a Phase II application, which is onerous, in order to get inspected and authority to operate.

Stansberry Research: And as of early 2018, that's still ongoing?

Lang: That's still ongoing.

Stansberry Research: What was it like to raise capital for this venture? How did it compare with other businesses you've started?

Lang: I'm an entrepreneur. If you look at my history, I've started four or five companies. I hate to be boorish about this, but to me, this is just starting another company.

I will say investors show a lot more interest in this cannabis company. All the other ones, I had to go work hard to raise capital. Hard. And this industry is the total opposite. Friends call you up and say, "I see you got a license. Are you guys taking investors?"

From that point of view, there is excitement out there. But I can tell you that my partners are all business guys. And to us, it's just another business.

The original partners put up the money to secure the license. After we secured the license, we did a round of financing and brought additional local investors in. They all lived in the Maryland area. And now that we're opening, we're going to do one more round to facilitate the operations.

Stansberry Research: So it's fair to say that everyday investors who live in a state where medical marijuana is becoming policy could have opportunities to invest with friends and families who open operations like this and see a decent return on their money over the next 10 years?

Lang: Completely.

Stansberry Research: I think one thing that holds up a lot of folks from taking the dive into cannabis investments is they feel it is still a business dominated by nefarious characters. Has that been your experience at all?

Lang: I've not seen a hair of that.

I'm pretty plugged in here in Maryland. I don't see any of that. It goes back to the arduous process I mentioned earlier that's required before you can even get your foot in the door as an entrepreneur in this industry.

When you supply this information to the state regulators, I don't see any chance for nefarious characters to get involved... You have to submit to background checks, fingerprint checks, FBI checks, five years of tax returns, and bank statements. I don't know how a slimy guy would slip through that.

I don't see much change coming to the regulatory framework that's been set up. So I don't see that [process] changing in states like Maryland. I think Maryland has done a good job. It was modeled after Arizona, which was launched a few years prior to Maryland.

And the way the medical framework has been set up in Maryland and Arizona is a lot more disciplined than other states. Even so, from a competitive standpoint, the model for the industry in Maryland is good...

There are a total of 15 growers, so you're going to have **a lot of competition** in the growers market. Of course, these are all vetted and financially adequate, sound companies. So you're going to have **a lot of supply**. There's also going to be a total of 104 dispensaries. That sets the stage for **a lot of competition**. Each senatorial district has two dispensaries. But it also is protected turf. So each senatorial district has two, but only has two. *From that perspective, the business model is good and favorable to the investors behind these operations.*

And I don't think any of that is going to change anytime fast. We go to all the commission meetings or we dial into them. Early on, maybe a year ago, they were talking about changes. I don't even hear talk about changes anymore.

Stansberry Research: Do you think one of the benefits of having a model like we do here in Maryland is that the regulatory aspect provides a level of security in terms of product quality and consistency... in terms of what people are getting?

Lang: Absolutely.

This was a little hard for me to switch mental gears, when I started going through this process. In Maryland, you really have to think of this as medicine. Thinking of it as what you normally think of it as, that isn't going to help you analyze or run a business.

You've got to think of it as a medicine. You've got to start a business that's selling a medicine.

Stansberry Research: In other words, investors should think of the medical cannabis business more like they think of biotechnology start-ups?

Lang: Any pharmaceutical. It's like a pharmaceutical company. All the regulations, all the procedures... You have to have patience. [Your customers] have to have seen a doctor. They have to be registered with the state. You're selling a medicine.

If you want to invest in companies in the medical space, either privately or through the public markets, you need to approach your decision as though you were investing in a small pharmaceutical company.

Stansberry Research: Right. I guess the difference being that you don't have to worry about pot passing stage three clinical trials, right?

Lang: That's all been done in the '60s and '70s!

Stansberry Research: So if you expect things to stay the same more or less going forward in Maryland, do you see a time in the future when recreational use will be opened up on the federal level?

Lang: I think that's unlikely at this point.

Stansberry Research: Will that have any impact on the passage of recreational-use regulations at the state level?

Lang: I think the potential tax revenue for the states is irresistible.

Stansberry Research: Irresistible. Yes. I tend to agree.

Lang: And by the way, medical is not taxed. So states that just have medical programs don't derive significant tax revenue from the industry.

Stansberry Research: I'm glad you brought that up. Readers who aren't as familiar with the space might wonder what is the benefit for states that open up medicinal dispensaries?

Lang: Well, it's like I keep saying... And this is something I want your readers to understand...

I have not been a user of cannabis for many, many years. Of course, back in college, everybody did. But it's just not something that appeals to me. As I mentioned earlier, my education was in chemistry, and I started my career selling pharmaceuticals.

Over the years, I followed the medical studies on cannabis, which is scant. There aren't a lot of medical studies or evidence of the efficacy of cannabis applications, mainly because the U.S. Drug Enforcement Agency (DEA) has made it against the law for most universities to do any medical studies with THC, with cannabis. But there are studies, mostly from overseas, that show how powerful cannabis and its byproducts can be when used as a treatment for a variety of illnesses. And there are a range of applications. I think Maryland wound up having 12 indications that can be treated.

Here's what I mean...

The growers each grow particular strains. And they have to list the strains they're growing with regulators. Every time they harvest and do a shipment of cannabis, they have to send it to a Maryland-certified testing lab.

At the test lab, they have to have each strain tested for the molecular makeup of the strain... how much THC, how many CBDs, and so on. That's the point of view from which this is really a medicine. When somebody comes in and needs treatment for chronic pain, they'll use one ratio of chemicals in the cannabis. Someone with post-traumatic stress disorder will use more CBDs and less THC, maybe.

Consider processed goods, like oils and tinctures and edibles. They make up something like more than 50% of the market. Patients are seeking out cannabis in these forms because you get a very precise dosage. And by the way, a processed good can be a tablet. So you can buy a jar of THC tablets in a dispensary. But the point is, you can get a very precise dosage of medicine.

Now I know a lot of people simply aren't willing to believe that cannabis can be used for legitimate healing purposes. They think the entire medical side of cannabis is a ploy to get recreational users access to the drug. But that's not at all the case. It's actually a legitimate effort to get medicine in the hands of people who need it. The fact that there is a lot of money to be made is obviously a big bonus.

Stansberry Research: So at this point, you'd say this is a legitimate area of the markets for investors to consider allocating money to?

Lang: There's no doubt about that. Not in my mind.

Stansberry Research: The last few decades have seen huge disruptions across virtually every industry... from technology to entertainment, and most obviously retail... Is that something you need to keep in mind when investing in cannabis companies?

Lang: When it comes to medical cannabis, I'd say no. As I mentioned earlier, in Maryland as an example, only two dispensaries are allowed in each senatorial district. So there's a built-in competitive edge from that perspective, and I don't expect the model or regulatory framework to be impacted by any kind of disruption.

From the recreational perspective, I think the big disruptor to worry about is the DEA.

Stansberry Research: That's a good point we haven't really covered. What do you think about the political risk posed to the industry by a Republican-controlled Congress? Furthermore, what should investors be thinking in light of Attorney General Jeff Sessions' January 2018 comments about federal enforcement?

Lang: My personal take is that there are too many states hooked on the tax revenue. I mean, what is the federal government going to tell those states? "Sorry, Oregon. We'll replace the $100 million you're not going to get next year."

Of course, that might sound unbelievable if you've read the official letter Jeff Sessions wrote where he declared his intent to enforce federal laws prohibiting cannabis.

But here's why I think what Sessions did makes a lot of sense... Trump and Sessions' No. 1 priority in terms of policy going into 2018 is illegal immigration. Illegal immigration was Trump's top campaign promise, and it remains the most important issue for Trump's supporters.

Right now, the big roadblock for Trump's illegal immigration policies is states' rights. Just look at the sanctuary city debate. California is a good example of a place where states' rights are preventing Trump from achieving his goals.

Likewise, a big focus of Trump's immigration crackdown is getting rid of violent criminals... and Trump has extensively named MS-13, the gang, as one of his big targets. But the Obama-era letter got in the way of prosecuting gangs in the drug trade.

Now keep in mind that Trump has historically been a big states'-rights guy. So he can't be seen as taking away powers from the state. Which is why in the past, he's always maintained that cannabis is a matter better left up to the states. During a campaign rally in 2015, he was famously reported as saying that he thought cannabis was a state-by-state issue. About a year later, he also said he was in favor of medical marijuana 100%.

So when you think about it, Sessions' letter is actually brilliant in the context of Trump's goals and challenges.

Trump and Sessions want to crack down on sanctuary cities and gangs made up primarily of illegal immigrants, but Trump can't be seen as an anti-states' rights guy. So he uses Sessions to enforce the federal government's oversight. That way, they can use federal officers to break up sanctuary cities and use federal policy to convict violent gang members for possession of marijuana in states where laws don't prohibit it.

But I don't think he will actually interfere with any state laws that might conflict with the feds. If a state has a legal framework for medical cannabis, I think he leaves that alone. They will go after activity that is illegal under both state and federal laws without a doubt.

Trump trumps Sessions at the end of the day. So this is sort of a win-win politically for Trump. He gets his cake and eats it, too. That's my personal opinion, of course. But I don't see it being a big threat to any cannabis businesses operating within a legal state framework.

Stansberry Research: Where do you see the industry a few years from now? What does it look like down the road from here?

Lang: I would say that the reason you invest now is because ultimately, I think Big Pharma is who buys up everything that's out there now. If not Big Pharma, maybe a consumer-products company.

Stansberry Research: So you see in the future a huge wave of consolidation, of these mom-and-pop shops being bought up, possibly by multinationals...

Lang: The only thing that could stand in the way of that is state regulations that prohibit cross ownership.

Stansberry Research: Like you see in alcohol distribution.

Lang: Like with alcohol. But you don't see that with medicines. And at the end of the day, I do think Big Pharma will end up dominating this industry. That might be 10 years from now, but the revenue potential for them is going to be irresistible.

Stansberry Research: Yes, it's not often in this lifetime that a completely new commodity suddenly appears in the market, like cannabis...

Lang: You mean like bitcoin?

Stansberry Research: Well, it seems we're getting a few more examples recently than normal.

Lang: Markets certainly haven't been boring.

Stansberry Research: In terms of investing, should folks think of what's happening with cannabis like the gold rush? Are the "picks and shovels" going to do the best?

Lang: I don't think it's much different than any market, like semiconductors in the 1970s.

[Cannabis] is going to be a rapidly growing market for 10 or 15 years. So you can pretty much pick any area, as long as you don't get with a shady operator, and it's going to get swept along.

If you have a lot of capital, you probably want to look at growers because the growers are consuming the most capital. **If you just have a little bit of capital, you want to look at being a limited partner in one of the dispensary operators**. But there's opportunities for any kind of investor who's got some means.

Stansberry Research: In 2017, we saw Scotts Miracle-Gro (SMG) stock soar when the company acknowledged publicly that it was actively doing business with marijuana growers. Do you think right now people should be looking for opportunities like that – established businesses that are going to participate in the marijuana boom? Or do you think some over-the-counter stocks are worth considering?

Lang: As with everything in investing, there's no getting away from risk/reward ratio. Miracle-Gro is going to be way less volatile than some four-year-old outfit in Colorado that started making its own brand of edibles and is trying to set up distribution operations in states that have legalized medical cannabis.

From an investing standpoint, there's the total spectrum in this industry already. Everything from private investment... to really small startups that are branding their own stuff and building distribution networks... to established, billion-dollar outfits that are trying to develop a niche in the greenhouse space for growers.

I don't follow many of the publicly traded entities in this space. But what I can say for sure is that *private investors have plenty of opportunities to get some skin in the game*. And if you can

get invested in the industry through private means, not public means, you stand to do even better than you would buying into any publicly traded stocks at the moment... and you don't have to tolerate the enormous volatility of the public markets, either.

As far as the equity markets go, the Canadian markets are more developed, equity-wise, than ours. That's probably where you can do the best-quality research.

What's been a good area of investment is investing in companies that are building state or national brands and processed products. Outside of the medical cannabis universe, for many of these companies, cannabis is going to be commoditized. As more people grow cannabis and there's more competition, the prices for cannabis will eventually decline – either a little or a lot.

So the companies that are building branded products are going to have a competitive moat that say, a grower in California, might not have. Even circling back to medical cannabis, what the consumer wants is a dependable product that he knows, trusts, and can reliably expect to be the same as the last time he used it.

So if you are looking to get exposure to the cannabis industry, it's worth taking a look at if the entity you are investing in produces processed products or at least gets its revenues from its product mix.

Stansberry Research: What do you see as the biggest dangers for investors in marijuana companies that they should consider?

Lang: The public companies? I don't think it's any different [than any other industry].

I would look hard at the management team. Make sure you've got a solid group there. There's tons of those out there. I'm sure you can find 30 companies with solid management teams.

Then look at the product and product strategies, like you do with anybody. If you see a solid management team that has established a branded product in Colorado, Washington state, and Arizona... Buy.

Buy. Because five years from now, that branded product will be in 40 states.

Stansberry Research: Do you think that in the event of any broad market decline or stock market decline, these businesses will be in jeopardy?

Lang: Well, at least historically, we've seen that some industries are insulated from recessions. Gambling. Alcohol.

Stansberry Research: Pharmaceuticals?

Lang: Right, pharmaceuticals. People still take their diabetes medicine, even though there's a recession.

But there is an important difference...

The cannabis businesses could be hurt more than pharmaceuticals during times of volatility. That's because this is a cash business. There's no insurance reimbursement. There's no processing with credit cards. It's probably the hardest part about being an entrepreneur in this industry.

And that's the case in every state because of the DEA. No bank that has any kind of federal oversight, like the FDIC or U.S. Treasury, will do business with any company in this area because it jeopardizes their relationship with the federal government. So your cash-management part of this business is still a big challenge.

By the way, that's an area I'd be really interested in as an investor – trying to find companies that are building solutions to cash-management challenges in this business. I suspect a couple companies are going to grow up to be huge businesses if they focus on that problem.

We have been contacted by a couple companies that have started cash-management operations inside of a state. So they escape the federal regulatory scheme. But we're only talking to them. It's too early to say that they have a product we're going to use yet or not. But regardless, that is a big costly problem that is waiting for a solution.

Here in Maryland, there's only one bank I know of that's gone through all the regulatory hurdles to get approval to manage cash for this business. I don't know if you can invest in it, and I'd rather not say the name publicly. But I would look for opportunities like that. The company has nothing to do with growing or dispensing cannabis. It's an ancillary business that's going to wind up making a lot of money out of this industry because it solves a real problem for the industry.

So there's all sorts of places you can look... a virtually endless stream of opportunities that all have the potential to be big winners over the next decade.

– Chapter 3 –

The Future of Legal Marijuana

A conversation with Matt McCall
Founder, Penn Financial Group
Editor, NexGen newsletter franchise

February 7, 2018

Matt McCall is the founder and president of Penn Financial Group – an investment bank designed to guide its clients toward financial freedom. He is also editor of the NexGen newsletter franchise – published by our corporate affiliate InvestorPlace Media.

Matt is known across Wall Street for extensive top-down technical analysis that leads to precise buy and sell points in the best stocks and exchange-traded funds (ETFs) that pass his screenings.

Across his multiple newsletters, more than 80% of his recommendations since 2015 have been profitable for subscribers.

A highly sought-after speaker in the financial industry, Matt has made 1,500 television appearances in the last 10 years. He served as the chief technical analyst and co-host of national radio show *Winning on Wall Street*, co-hosted daily investment show *Making Money with Charles Payne* on Fox Business Network, and contributed to the Fox News Channel.

Starting in 2010, Matt was one of the first analysts in America to cover over-the-counter marijuana stocks. He became a staple figure at marijuana investing conferences, panels, and events. Matt is on a first-name basis with the CEOs of the world's top publicly traded marijuana companies.

Matt is a staunch advocate of not just telling people what to invest in... but educating people so they understand exactly how these investments work for them.

He is also the author of two books on investing: *The Swing Trader's Bible: Strategies to Profit from Market Volatility* and *The Next Great Bull Market: How to Pick Winning Stocks and Sectors in the New Global Economy* (a top-selling investment book for more than two years).

In the following interview, Matt discusses openly and honestly his views on the future of the legal marijuana industry, the headwinds it will face, and the eventuality investors should prepare for, including...

- The simple reason he expects cannabis will become legal in every U.S. state.

- How to structure a cannabis portfolio in 2018.

- The timeline for federal legalization of medical and recreational marijuana.

- The biggest dangers investors should consider.

- And the one factor that will convince the world cannabis is a legitimate industry.

Stansberry Research: Hi, Matt. As an equity analyst with a long career that includes a show on Fox Business, what first interested you in looking at the marijuana market? When did you start educating yourself on this industry?

Matt McCall: Seven or eight years ago, I asked myself, "What industry out there has the potential to grow to a multibillion-dollar industry outside of just technology?" Marijuana is the one that kept coming into my mind.

At the time, cannabis was just starting to catch on medically. I started seeing some of these penny stocks creep up. And I started getting a lot of inquiries from different people about these stocks. Everyone who followed my work knew I always like to be on the cutting edge of what the next big industry is going to be.

So I started seeing that happening and, just out of curiosity, started doing a little bit of a deeper dive into the companies. What I realized, at that point, was there was a complete lack of credible research and information available on these opportunities. I'd say 99% of the "research" that was out there was somebody pumping up a stock through marketing and then getting out of it.

It's similar to what we're seeing with bitcoin now. You were a biotech company last week. But when you change your name to "Blockchain," your stock goes up 1,000%. The same happened with cannabis companies over the last decade during certain periods. I realized the majority of the companies out there were junk. That was obviously disappointing to me. But in the years since, things have gone from the Wild West to not-so-Wild West. And some of the bad actors have been thrown out of the industry.

From the standpoint of stock analysis, you're seeing a huge demographic shift. The older demographic is dying off.

And a lot of people I know – in all walks of life, all ages of life, whether they make millions and millions of dollars or if they are living on minimum wage – smoke marijuana. They may not talk about it, but even that's starting to change.

Honestly, at first, I was just doing a lot of writing on the area and began publishing stuff and throwing it on the Internet. Then about six years ago, I started getting requests to sit on panels and moderate marijuana-investing panels. I've probably sat on 50 different panels with CEOs of all the major companies. From Canopy Growth in Canada – now a C$5 billion company – to the little guys that are now probably back in their parents' basement.

I was lucky to be in a position to see this entire industry morph from what it was to what it is now.

From a personal standpoint, I was in the hospital with a friend maybe four years ago – emergency room. There was a little girl next to me, maybe 10 years old, crying and crying. And the father was crying.

It was so disheartening, it was almost making me cry. I talked to the father for a minute and I said, "If you don't mind, what's going on?" He said, "She has terrible seizures and the only thing that helps her is cannabis." And I saw that firsthand. I thought, "Wow, this girl's life will be horrible forever unless she has that."

It's not about people getting high. It's the fact that cannabis is a true medical breakthrough that this country is so backwards on and still considers more dangerous than cocaine.

Stansberry Research: Yeah, it's a Schedule I narcotic.

McCall: I'll be honest with you, it's ridiculous. I mean, that is just insane to me. Alcohol is more dangerous than marijuana, in my opinion. But back to your question...

I saw how things are going to change, how these older people who have one point of view are disappearing. And then you have millennials and Gen Xers like myself who have largely more progressive views on cannabis. Even the Baby Boomers like my parents, who experimented with it 40 years ago and may still indulge, can see the upside potential. There's just so much money.

At the end of day, I like money. And we can make so much money in this industry from so many different angles.

I went to graduate school in Colorado, had my first job, and started my first company there. When I moved there, they had just started to open up gambling in Central City in that area. And I remember getting money back from the government.

I was like, "What the hell's this?" They made so much money in taxes that we got money back. So Colorado was at the forefront of gambling. The same is true with cannabis. You now see gambling in every state basically. I think the same thing is going to happen with cannabis.

As of late 2017, 29 states and D.C. have legalized medical cannabis. Seeing how much money's going to be made will then encourage [the other] states and local municipalities to push for legalization.

I think within five years, it will be legal federally. And all these companies that are sitting at a couple hundred-million-dollar market cap will be several-billion-dollar companies, just like in Canada.

Stansberry Research: You said that over the last seven years, you watched the industry go from the Wild West to a little less wild. Do you see over-the-counter equities as being legitimate considerations for investors right now? Or is that still too wild? And if so, how do you separate which are the good ones and which aren't?

McCall: So to go back to your original question, is there opportunity? I think there's huge opportunity here in two parts.

One huge opportunity, for lack of better words, is finding the unicorns that are really going to flourish once this becomes a federally legal business.

The second huge opportunity is to trade them because you can plan ahead.

California's recreational-use law just went into effect in January 2018. A lot of people bought ahead of the news. But once we see how much money is actually generated in California, it's going to be a boom for a lot of these companies. The companies that own dispensaries there. The ones that own the grow facilities there.

That's going to be a huge boom you can trade around. You can really play that by starting to buy in advance. Once the numbers start rolling out, these stocks will soar.

Then there's that longer-term play I mentioned. That's all about looking for the stocks that will stay around. And Canada is a great example. The government stepped in to allow Canadians to buy their medical marijuana online. They're looking at July 2018 to be able to buy recreational marijuana online and have it delivered to your home. That's another event you can trade around.

It's all regulated by the government, which is a great thing. Even though I hate governments, it's a good thing for this industry. And we've seen that in the results of those stocks and the increased perception of the legitimacy of this business. When alcoholic-beverage company Constellation Brands came in and invested $190 million in Canopy, that sent a big message to the world. Canopy now has a C$5 billion valuation. This is a serious, legitimate business we're talking about.

Constellation owns Corona. This is not just some fly-by-night company. This is a real company that's getting into the marijuana business. And that just shows that big multinationals are starting to position themselves for the moment the U.S. becomes legal so they can jump right in.

Now jumping to your current question about separating the good companies from the bad...

When I analyze these stocks, a lot of them get cut right away. First, I look at management and try to see where the people come from. Was this company a biotech two years ago that changed its name and said it was a cannabis company overnight?

The second thing I do is look and see if they're actually making money. Some of these companies have no money. They may have so many shares outstanding that it's basically just like this scheme of people trying to pump and dump with press releases.

Stansberry Research: Of course. Remember the marijuana vending machine company Medbox? Remember how much it was trading for a few years ago? It completely tanked and the CEO was sued.

McCall: I know, it's ridiculous. I know these guys. I know the bad players personally. I'm not friendly with them, but I know them and I know what they do. They make millions of dollars knowingly preying on the poor American public. And they're great at it.

In short, you are going to end up analyzing these companies the same way you would analyze any company and any industry. You just have to have a different mindset because news is very important in this industry.

Around election time, you're going to probably get a spike. Usually coming into a new year, you'll see a spike, like we saw in January 2018 when California opened up. I expect you'll see a spike around when Canada opens up around July 2018.

If you're a trader, try to trade around it. And then buy into the dips.

Stansberry Research: Is it legal to buy over-the-counter stocks? Do I not have to worry about anything, as long as I pay my taxes? Or is the government going to come after me?

McCall: No, you're fine. You don't have to go to TD Ameritrade with a bag full of cash. You're OK.

Stansberry Research: I know your expertise is in trend analysis. Do you apply that trend analysis to trading pot stocks and have you had success with that?

McCall: Yes, I have. We did a little report on it back in November 2016. Played the stocks for a pop. Some of them we held for six or seven days and got good 20%-30% gains in a short amount of time.

I haven't done as much recently only because there's kind of this stigma around it. I didn't want to be typecast as a "cannabis guy" because there are other industries and markets that I think offer terrific upside as well.

GW Pharmaceuticals has been a way we've been playing that for a while now on and off. It's been doing great. Hitting new, all-time highs in early 2018.

I have a long watch list that I go through every day. But I'm very careful because I don't want anyone to get burned. **This can be a**

volatile industry. I don't want to be that guy who's pushing these stocks in a bad way.

But at the same time, there's opportunity. And I don't want folks to miss out, either. I also worry if they don't come to me, they're going to go to somebody else who's almost certainly trying to fleece them.

Stansberry Research: Right.

McCall: So that's why I'm honest about it. These are two ways to play it... Either you let it go and live through the ups and downs, or you trade it. There are risks in both.

Stansberry Research: You've been watching this market now for seven years. Do you think that we're still in the early stages? What inning are we in right now?

McCall: Second.

Stansberry Research: You think we're in the second inning.

McCall: I put Canada in the fifth inning and put us in the second inning. I'm a numbers geek. And at the end of the day, it all comes down to how much money you have in revenue and earnings.

Just look at some of the numbers... Canada, for example, is expected to generate between C$2.5 billion and C$4.5 billion by 2021. That's great for Canada.

However, recreational in California alone is projected to be greater than Canada by that point.

Stansberry Research: Have you looked a lot at Canadian stocks? Do you think that's a place people should be looking to for more established, legitimate equities than what generally make up the over-the-counter markets here in the U.S.?

McCall: Yes, I do. Today.

Stansberry Research: Today?

McCall: If you say to yourself, "I'm going to put just an arbitrary

10% of my portfolio into cannabis stocks," then you want to do 5% and 5%. You want to have some in Canada because those are the companies that actually have revenue. Some of them have earnings, believe it or not, and a true business model. And they don't deal with any legal issues because it's all run by the government.

So that's a lot of risk they're taking away. That can become your core portfolio, and the more speculative part of your portfolio would be the U.S. stocks.

Also, a lot of the Canadian companies are expanding into Australia and other areas to start growing revenue internationally. So it's a global play as well.

Canada also has an ETF: Horizons Marijuana Life Sciences Index Fund (HMMJ.TO). It launched April 5, 2017. Since that time, it has grown to something like C$310 million in assets.

Then there's the first pot ETF in America: the ETFMG Alternative Harvest Fund (MJX), which focuses on marijuana-related companies.

MJX launched on December 26, 2017. When I looked at the ETF the day it commenced trading, it had $5 million in assets under management. One month later, it had $418 million in assets.

The acceptance by large investors shows the upside potential in the marijuana sector.

Stansberry Research: And you think that's a legitimate ETF to consider?

McCall: Yes. We recommended it in my NexGen newsletters a couple days after it began trading. In the first month after the recommendation, the ETF was up 26%. We're happy with the holdings of the ETF as they are spread between both Canadian and U.S. marijuana stocks.

We liked some of the holdings so much, we also recommended two stocks in the newsletter. One we hold with a 40% gain. The other we sold in less than 48 hours for a gain of 49%. This was one of those trading opportunities we were able to identify and profit from.

Stansberry Research: You think we'll have legalization on a federal level in five years. I assume you're talking medical, correct?

McCall: Yes.

Stansberry Research: So what's your ballpark timeline for recreational? Or do you even think that that'll happen at the federal level?

McCall: I think it will happen eventually. You're probably looking closer to eight to 10 years down the road until the demographic shift is fully realized. It also depends on who's in office.

I'm in Nashville now, and Nashville is very liberal. But the state's very much Republican.

A friend of mine is trying to push through a bill to legalize medical marijuana here in Tennessee. He's getting pushback from these old, stodgy Republicans out in the woods that don't like alcohol, let alone marijuana. When they realize that their voter base is in favor of it, that will really create change. Because now you can get elected.

And I believe this will become a central topic for elections during the next presidential election. Medically, I think it could be really quick. That could be a hot topic for whoever the next president is to say that we're going to federally legalize it medically. For sure.

Stansberry Research: Do you think the FDA will sooner reclassify the status of marijuana as a narcotic?

McCall: It's going to be a bunch of little steps like that. And every time that happens, it puts marijuana companies back in the news and gets people excited again.

In the investing world, everyone is focused on bitcoin now. People have not realized how well the pot stocks have done. (They've done fantastic.) But they're under the radar because everybody's focused on bitcoin.

Stansberry Research: What do you see as the biggest dangers that investors should consider?

McCall: Well, one is not doing their due diligence... just like any stock, but even more in this space, because it's an industry that is illegal federally. It is not regulated in the same way as other things, so you need to go above and beyond in your research to make sure that you're in the right stock.

Also, too many people are only seeing the get-rich-quick opportunity with pot stocks and just go after a name. They go after something they heard at the water cooler, some baloney press releases they read online. If you're doing that, you're gambling. You're better off at blackjack.

If you're truly going to invest in this and make this a part of your portfolio, you have to have a plan and have a risk strategy as well. Let's take a 20% loss and get out. You have to view this the same way you'd look at the automobile industry, like if you're going between GM, Fiat, and Ford.

You need to be able to look and analyze it the same exact way. But people don't do that. They just think, "If it's related to marijuana, it will boom. It's going to go up." It's the exact opposite.

What has changed in the last seven years is that a lot of the bad players have been banished. The industry has started to self-regulate. You're seeing big money come in this industry, and that will push out the bad players and the fake people that try to capitalize on the name of cannabis.

You need to do your research. If you do your research, you're going to make a lot of money. There aren't that many industries out there that have the potential to grow as quickly as marijuana does in the United States and globally.

Stansberry Research: Earlier, you mentioned Constellation. Where do you see the big opportunities to buy into big-name companies that could logically move into this space in the future? Or maybe "picks and shovel" plays people might not know about?

McCall: There's AeroGrow International (AERO). The stock hasn't done much of late, but I think Scotts Miracle-Gro eventually buys it out at like $4 a share. It's at less than $3 in early 2018. So that could be a winner.

Anything to do with lighting as well. There's a player over in Europe. I've met the CEO a few times and really like him.

Look for companies that don't actually touch the plant, so they don't have to deal with the regulatory issues. There's another company called MassRoots. I know the CEO well. I've known him since he started the company.

MassRoots is like a Facebook for pot users. And it has over a million users. Nice business model. It basically shares stories about pot and strains and all that kind of stuff. I don't get it, but I see the potential.

MassRoots applied to the Nasdaq and got turned down because it was "illegal." That's BS to me. MassRoots doesn't touch the plants. It's doing nothing illegal. It's only in states where pot is legal. It can't have users in states where pot's illegal.

It followed the rules and everything. But that just shows that the Nasdaq wasn't ready for it. So look for a company like that, too, where once it can become listed, the stock will soar.

You could also consider Innovative Industrial Properties (IIPR). That's a real estate investment trust that owns growing facilities – another way to play the boom as well without touching companies that touch the plant.

The picks-and-shovels secondary space is great because those stocks can actually have a bank account.

So those are a lot of the different ones. And then there's some of the vaporizer companies. I met the Kush Bottles (KSHB) CEO a few times. He seems legitimate to me.

[Not all the secondary players are] going to do well. It's going to end up being the big names. And in this business, if you're going to be an investor in this for the next, let's say, three to 10 years or anything like that, you will see the big names come in and eventually run it.

Stansberry Research: That's what you see long term... a big wave of consolidation across the industry?

McCall: Absolutely. And you will see big names come in, too.

It will happen sooner than anyone thinks because Constellation was a big move. It spent a lot of money – $190 million. Now Budweiser and everybody else is sitting back thinking, "Heck, I could take a couple hundred million and invest in this space. That's nothing in our balance sheet."

Stansberry Research: Drop in the bucket.

McCall: Once we see two or three more big names move into the space and make an acquisition, that will tell the world this is legitimate. And that's going to happen in 2018. We're going to see a couple big names come in.

Stansberry Research: What are your thoughts on Jeff Sessions' recent official statement about cracking down on the marijuana industry? Does that change the picture for you?

McCall: What the Obama administration did in the past was adopt a "hands off" approach when it comes to enforcing marijuana laws with something called the Cole Memo.

When you consider that more than half of the states already had some form of legal marijuana in 2013, it was the right move by the feds to leave enforcement to the states.

That's the same reason why it was a terrible move by Sessions to rescind the Cole Memo.

I tweeted out that day that the move by Sessions was political suicide. Definitely for him. If the GOP backs him, it will be for them as well. Now that we've had some time to watch things play out, it appears Sessions has alienated himself even more than I anticipated and that his likelihood of making it through the entire Trump reign is low.

I want to explain two big reasons why Sessions is now on an island...

First, there are too many GOP-led states that are generating tax revenue from either medical or recreational marijuana sales that will not back his decision. Second, with the majority of voters now favoring the legalization of marijuana and the young vote overwhelmingly in favor of the move, politicians up for re-election will be forced to either back legalization or ignore the topic.

Remember, there are just two things politicians care about: re-election and money.

Washington D.C., along with the 29 states that allow medical or full recreational marijuana, make up nearly 63% of the American population. If you include the 15 states with CBD-only laws, 95% of the population lives in a state with some form of legal cannabis. Like I said, political suicide.

From the perspective of investing, the dips caused by anything related to Sessions are buying opportunities. If he authorizes the feds to raid a business that is legal under state laws, it would obviously be bad for the stock. However, it may be just what the industry needs long term. Federal raids making headlines would create a groundswell of support for legalization.

Either way, Sessions is a dinosaur that is out of touch with reality. And his decision will not stop the inevitable – legalization of marijuana at the federal level. His demise is around the corner.

Even Trump is a states' rights guy.

Stansberry Research: Is there anything we haven't covered or anything that you would want to know if you knew nothing about this industry and were interested in getting into it for the first time?

McCall: Well, again, I'm a numbers guy. And that's the only way stocks move higher. So look at Canada... As I mentioned, it's only looking at something between C$2.5 billion and C$4.5 billion by 2021 in recreational sales.

But the same study shows between $7 billion and $10.5 billion in 2021 for just recreational in the U.S., not including medical. And that's just a few states.

Imagine the size of how big this could be. You're looking at $50 billion-plus in just recreational. You add in medical. This is an enormous industry. Imagine buying into alcohol during Prohibition.

You give me other industries where you can say you're in a second inning. It's very tough to find that. The "Internet of Things" is in the fourth or fifth inning. And where is the money made? First three innings. You can still make money out there, but the big money is made during the first three innings. That's when you want to be in.

You have to take a little bit of a risk, obviously, but trends are getting better. The technicals haven't taken a step back, which is, to me, very promising as well.

– Chapter 4 –

The No. 1 Factor for Success in the Cannabis Industry

An interview with Doug Esposito
Cannabis Practice Leader, Owen-Dunn Insurance Services

December 27, 2017

Doug Esposito has been an industry leader in multi-faceted insurance programs and risk management since 1998. Today, he leads his firm's energy and cannabis practice with specific expertise in alternative risk management for these industries.

His current cannabis practice clients include indoor/outdoor cultivators... manufacturers, extraction, and distribution companies... and dispensaries and property owners. So he's familiar with the challenges facing the growing industry and is skilled at providing solutions.

Doug currently serves on the California Cannabis Industry Association's Insurance Committee as the head of public relations.

Doug has been an active member in the cannabis community for several years. He has been asked to speak at a variety of events, including the Sonoma County Growers Alliance, the California Cannabis Industry Association's Annual Policy Conference, the Insurance Business America's Cannabis Cover 2017 Conference, the Department of Insurance's Cannabis Public Hearing, and the California Cannabis Risk Symposium.

In the following interview, Doug shares...

- How something as simple as insurance can tell you everything you need to know about the health and future of the cannabis industry.

- How to know which investment opportunities in the cannabis space are viable, including niche areas.

- The No. 1 factor that will make cannabis investments soar in 2018.

- And how many billions of dollars could flow into banks when the cannabis "banking problem" is fixed.

Stansberry Research: Doug, thank you for taking the time to sit down with us today.

You work in the insurance business. That's a highly regulated industry. Founded in 1949, your company – Owen-Dunn Insurance Services – has a long history in the space. So you didn't just pop up to capitalize on this new market... You have an existing business that expanded into it.

When did you decide to move into the cannabis arena and why?

Doug Esposito: Our firm has been around quite some time. And you're right, the insurance space is highly regulated. It's also relatively conservative. Broadly speaking, insurance is akin to banking.

It all started for me personally a little over two years ago when one of my colleagues, a CPA, came to me and said legislative changes and compliance were changing in California.

I started doing my research...

As it turned out, my firm had insured several cannabis dispensaries and other related businesses back in the mid-'90s. However, most of those companies shut down. We weren't actively involved in the space anymore until October 2015 when the Medical Cannabis Regulation and Safety Act (MCRSA) passed.

I was looking to expand my portfolio of clients outside the construction and energy spaces where I primarily operate. Once you're in a space long enough, you really learn the technical sides and you get engrained, so I wanted something else to do in addition.

I live in Sacramento, the capital of California. A lot of the politics and the policy work was and still is happening here, so it puts me in a good spot to test everybody's pulse, get connected to the California Cannabis Industry Association (CCIA), and become active. Sacramento is also a good area from a cannabis policy climate.

Stansberry Research: It sounds like Owen-Dunn had previously insured a couple other businesses. But I imagine you still have significant regulatory challenges to overcome, right? Has that been a big part of your experience as a pioneer in this industry, or has that not really been a problem at all?

Esposito: From a regulatory standpoint, there haven't been too many hurdles. You see, the way insurance is regulated is actually state specific...

The federal government gives the states the ability to manage insurance on a state level. The states can say, "Hey, that policy is valid and legitimate in this state and this could be enforceable." That's how we've been able to provide insurance coverage despite the fact that cannabis is obviously still federally illegal.

The challenge is not as much the legislative piece on the insurance... it's the heavy fear that the feds are omnipotent. There is always the risk that they shut down an operation and seize everything despite the fact that the business is operating in accordance with all state laws.

So one of the big challenges from an insurance perspective is quantifying that risk.

As a result, there just aren't a lot of carriers that will play in the cannabis space. There are basically two legitimate resources that sell real coverage for virtually every line of insurance, like product liability. Most firms don't want to touch product liability. If someone gets sick or injured from smoking the flower, there are only two carriers that truly are not excluding that when you get into the 80th page of a 100-page policy, you know what I mean?

The other challenge from an insurance perspective that is obviously tremendously impactful is banking.

Stansberry Research: I'm glad you brought that up. As I understand, from a practical standpoint, these are by and large all-cash businesses, correct? So how does that work from an insurance perspective?

Esposito: It's very challenging. The bottom line is if you want to work with enough people and you're going to write their insurance, you need to collect cash because that's what they have. Some of our premium dollars we collect in cash.

Stansberry Research: That's incredible.

Esposito: We just step right in line with the industry. One of my colleagues said they were at one of the dispensaries, and it was payday. People just lined up like a soup kitchen. But the difference is that they weren't collecting soup... They were collecting rolls of cash for their week's wages.

From an insurance standpoint, that really scares me on a lot of levels because without banking, you've got so much more theft and so much more opportunity for money to be stolen and laundered.

Now the state has begun mandating that you have third-party security and that is a step in the right direction from a safety standpoint. It's just a huge pain because the industry is forced to manage picking up cash, which isn't convenient. It's a tremendous waste of time and resources.

It's really crazy to consider how big California's market is now with our recent adult-use recreational legalization, and simply how much money is going through these businesses. The lack of a financial solution for that is truly mindboggling and a shame. But in time, it will happen.

Stansberry Research: Are there any banks right now that are working with these businesses?

Esposito: There are several savings and loans and credit unions in the space. But they're all quiet, and they don't advertise...

First, they don't want that exposure in case someone gets a little bit disgruntled and wants to make another example of them with federal enforcement. And second, they'll get inundated with cash because the cannabis companies are truly exhausted with the cat-and-mouse game of trying to get legitimate banking. There's far greater demand for banking to the cannabis industry than there is supply right now.

Plus, you have companies with accounts shut down every week that are constantly trying to open new ones. It's a constant source of frustration and wasted energy.

The way they get around it is by having complex corporate structures. From an insurance perspective, that makes my job keeping up with those structures tricky.

I want everybody on the policy so we can have the most protection in the event of a suit or claim. But these companies also structure it in a way where the person on their banks' accounts is one or two arms removed from actually touching the cannabis product. And that is their one kind of banking arm.

If you really think about it, the time and the energy to open up four or five different accounts a year and then to play that game is substantial. Added to that is the challenge of paying for taxes with cash because the professional people – the CPAs and the attorneys – don't want to handle the cash.

I don't either. I would prefer a wire transfer or a check every time. I'd pick it up as a convenience to them, but it is a notable risk.

Stansberry Research: Do you see any regulatory shifts coming down the pipeline that you think will change that situation?

Esposito: When I talk to the people who are on the association committees... on the state committees with the highest levels... with the treasurer, they say candidly behind the scenes: "No, we don't see a solution soon."

Stansberry Research: So what it comes down to for folks who are investing in cannabis companies in the United States is that until we see a change in the status of cannabis at a federal level, these companies won't have access to an FDIC-insured bank?

Esposito: That or, maybe more likely, it might take a year or two for the state to come up with its own solution and have it all backed by the state of California.

Stansberry Research: What's interesting is the sheer amount of money – taxes – that the government is making off of this. It's got to be difficult for the federal government to tell a state, "Hey, you can't take that extra $2 billion in taxes." And the same applies for banking. At some point, the banks won't be able to resist all the money they can be making from these businesses.

Esposito: Absolutely. A lot of it is the political fallout. It's the reputational brand fallout, meaning banks are worried they will receive negative press and their stock price could fall as a result if they are associated with the cannabis industry.

Here's what's funny... Think about the California banking system. The California Department of Tax and Fee Administration takes those taxes and puts them in the bank. Now it's going to have to staff up because it's going to have millions and billions of dollars in $20s. No kidding, 20-dollar bills. It's going to have to count them and recount them.

It's a huge task to have the whole checks and balances worked out for all the cash and deal with the constant problem of "sticky fingers" along the way.

It's going to be a mess in that respect. But here's what I think... Republican or Democrat, taxes are taxes. This type of unique subject matter will elevate itself above partisan politics before long... There are just too many benefits.

On the medical side, let's just take conservatives. How can they say no to the medical benefits of cannabis? Let's say they allow the type of testing that should be done and funded, and we collect credible data that proves some of the benefits we've seen in medical trials concerning psychotic spectrum disorders (PSD), seizures, and pain management...

Holy cow. The medical cost savings, not to mention how these medicines could ease people's lives could be tremendous. Improving people's quality of life is a bipartisan issue.

Stansberry Research: What's your ballpark timeline for when this happens?

Esposito: No way this is going to happen with Trump's current attorney general, right? And then I would think whether it's a Democrat or a Republican who comes in next, I'd give them two years to make it happen. So that's three years of the current administration and an additional two years for the next.

I'm maybe more optimistic than others. I've had people just look at me and go, "There's no way, Doug." And I say, "Come on. Once we get billions and billions pumping through California, they're going to have to do something. They really are."

So I'm going to go on record: I'm going to say within five years.

Stansberry Research: And if Trump serves a second term?

Esposito: I think Trump would reverse his position on the issue quickly if he is re-elected for a second term. At his core, he's a business person. And if it makes good business, he would have a difficult time not supporting it.

Stansberry Research: So Doug, you're in California which is probably the most developed market in the country. Do you think most of the money has already been made or do you think we're just seeing the tip of the iceberg? In other words, are we in the first inning or the last inning or somewhere in between?

Esposito: I think it's the tip of the iceberg. I've been going to many of the business conferences like the international business expo in San Francisco. I went to the National Cannabis Industry Association's business conference in Oakland, the CCIA's expo in Anaheim, and the MJ Business Conference in Las Vegas (the largest nationally). It looks and feels very early.

There are a lot of people who are positioned to be very successful. The question will be once the true heavyweights like Anheuser-Busch and the pharmaceutical companies really get in and play unencumbered... what will it look like?

That's the big question mark. But there are some really good people developing brands with some smart business managers. These are folks who have run $100 million and $200 million companies and see unprecedented opportunity in the cannabis space.

So I think growth is going to be tremendous from here. But it's like anything else...

There are so many people who are wonderful artisans but don't know how to run a business. They're having a tough time just trying to get into compliance with the local and state rules and licenses.

Without a doubt, a lot of people will go under, similar to the restaurant industry. I'll bet a third of the people go under, if not more. But there are some wonderful opportunities for people who do their homework, collaborate with talented colleagues, and work hard.

Stansberry Research: Is it fair to say that when you're underwriting these policies – just like in any other market – these are legitimate business people with legitimate backgrounds and credentials? Or would you say it's a lot of dreadlocked, pot-smoking hippies?

Esposito: No, the prior. If they obtain the local and state licenses and pass our underwriting criteria for insurance, they are in a much better position to be successful.

I can't take credit for that level of underwriting. The carriers that have decided to work in this niche have really put a lot of thought into their policies, procedures, and underwriting criteria. The insurance carriers and the client's interests should be in alignment. The carrier wants to help the grower be as safe and protected as possible. That benefits all parties.

In the cannabis industry, it's not the same as a conventional business. If there is a fire for example, and they don't have the proper insurance, they can't simply go get a loan from the bank down the street. Without the insurance, they are in trouble.

For example, let's say you're an indoor grower and you want me to insure your product. You've got your clones, your vegetative plants, and your flowering plants until harvest. When you harvest that product is when it's the most valuable. You're going to hang it and dry it and then sell it. But you might want us to insure the plant – which is an asset – for theft and fire from the time it sprouts until harvest.

We're not going to insure it for pollutants or mold because, let's face it, if you can't grow the plant properly, that's a business risk, not an insurable event.

You have to have a sprinkler system. If the facility isn't fully sprinklered, the cannabis has to be in one-hour-rated fireproof vaults or safes. And there have to be 24/7 surveillance cameras and motion sensors with a 14-day back up. It's really a high level of risk mitigation and not for the faint of heart or someone on a shoestring budget.

If the building is over 20 years old, you have to show proof of updates to the electrical, HVAC, and roofing systems.

The key point to understand is that the underwriting is very meticulous.

This is a challenging marketplace from an insurance standpoint. There are five "no's" that you typically never want to have in a general liability policy. A cannabis policy will have all five, like a "claims made" form, per-claim deductibles, defense is inside the limits, products liability is typically excluded, and coverage is found in a separate policy often with a separate carrier and a duty to defend.

But that doesn't mean the company isn't going to be profitable or that it doesn't have smart, driven folks behind the wheel. It simply means we need more insurance choices. And that is going to take time.

Just the basic requirements you must meet to get a local permit are extensive or massive. They're looking at your water usage, what you're doing about odor control, security, and your whole business plan. The city and local governments are really doing a deep dive to make sure these folks are looking at everything in an effort to further their success.

So if you're looking at investing in a company that has already made it through permitting and managed to have an insurance policy written, the company you are looking at is solid.

You can trust that those folks have spent some serious time and money in their statement of processes and procedures, getting off the ground, et cetera. And that's not easy. Trust in nothing. Do your homework. Review all the various documents needed for the permitting, especially the business plan and financials.

Stansberry Research: Doug, it is so rare that a completely new market opens up like this in someone's lifetime. How does it feel to be participating in a new market that 90% of America hasn't even woken up to yet?

Esposito: In that respect, it has a lot of allure to me because I love the underdog. I can't wait to see data come out that proves cannabis can be effective against a migraine, a headache, or seizures, et cetera.

It's exciting because there's some serious money to be made.

We're talking the whole spectrum. From outdoor growers that are producing a flower to pharmaceutical companies studying the technology for dermal absorption applications and the like.

And listen, let's say you just want to get high and relax. Instead of drinking a glass of wine or martini, you eat an edible. Or you vape some cannabis oil, the CBD/THC gets right into your bloodstream, and you feel relaxed and happy while you enjoy a nice dinner.

They're both mind-altering, and you can imagine a huge market for adult-use that could by itself be as big as the alcohol market one day.

Stansberry Research: How has it impacted your life and wealth as a whole?

Esposito: We're still in the early stages, so while there's huge opportunity here, the numbers are relatively small compared with my construction clients. But we're excited because we are positioned to be a front runner in coverages and an expert in this type of risk management.

In an industry with so many fast-growing companies, I could see the numbers far outpace my construction clients in the next few years, believe it or not.

Stansberry Research: While people may think a lot of the money is already made, there are so many gaps and opportunities in the market. It's amazing to think something as simple as insurance for this industry can be a big business. Average investors simply don't understand how many services and needs this industry lacks that are huge areas for potential investment and innovation and entrepreneurship.

Esposito: Yes, that's why it's so exciting. We have a client that's doing track-and-trace software – technology that monitors all the cannabis as it moves through the supply chain to make sure it isn't being diverted to the black market or to assist in recalls or product safety.

That type of software isn't new to this industry. All the products have to be monitored closely to prevent them from falling outside of the medical and adult-use cannabis pipelines. But this client just put their spin on it...

The technology is completely handheld in someone's cell phone and you can have a body camera that can also be another step of tracking and tracing when you have the video that goes along with the transfer of the barcoded product from A to B.

That's a basic idea that has become its own successful business. This client doesn't grow or sell any cannabis at all. It's essentially a technology play.

There are so many areas to add value and to be creative and to come up with a new spin or a twist – taking ideas and technology from tomato farming to cannabis or from beer distribution to cannabis. There are many crossovers.

I know people personally who are waiting for a lot of the extraction companies to go out of business so they can pick up the machines for $0.10 on the dollar. An extraction machine can cost anywhere from $100,000 to half-a-million dollars, depending on the volume of oil that it can move. It's real money, even in these more niche business areas.

Stansberry Research: There are so many ancillary businesses that most folks don't even consider... like the third party that tests the cannabis between grower and retail or the security services needed because it's an all-cash business currently or the businesses selling specialized fertilizer or grow equipment. And that's before you consider financial services and insurance.

Esposito: Absolutely. I think of it as birthing a 100-year-old baby. Cannabis has been in California forever. But it has operated as a black market for so long that the legitimate services and infrastructure required to make these businesses work don't fully exist yet.

I look at how I can bring my solar and energy-efficiency knowledge from the construction world into the [cannabis] companies we insure so they can lower the cost of goods and not get crushed when the cost of the flower goes down. Who's going to be more efficient and profitable to be able to weather the storm and keep moving as the landscape gets more competitive?

I love it. It's cool to see so much opportunity in these ancillary spaces.

Stansberry Research: If you had to say what's the No. 1 thing in 2018 that will make cannabis investments soar in value, what would it be?

Esposito: If California took more control and was more progressive on the banking solution, that would be a huge boon for the cannabis industry.

Stansberry Research: If they solved the banking problem, how much money do you think would flow into banks?

Esposito: Oh my... When I took my first insurance applications two-plus years ago, no one wanted to give me an address of their grow operation. So I'd tell them, "OK, this isn't going to work. I need to know your actual physical property address. I need to drop it in Google Earth and see the surrounding exposures... if the carriers will research the fire zone and flood zone... and if there's earthquake activity at your facility location."

That was just two-plus years ago. But many of these companies are slow to change because they are accustomed to operating below the radar as much as possible.

Stansberry Research: Are we talking $1 billion in income flowing or are we talking tens of billions or more?

Esposito: Tens of billions. They're saying that for 2020, the projection from some of those serious data sources in the space is north of $50 billion.

Stansberry Research: That's a lot of money.

Esposito: Yes, $50 billion annually. It's going to be a complete game-changer.

Beyond that, you can just imagine when interstate commerce is legal, what that could do for these companies. And now I'm going out on a limb, but let's say the feds drop their Prohibition...

A good example is an analytical lab. It can take anywhere from $250,000 to more than $1.5 million or more to get equipment in there to be able to really analyze the products for residues, solvents, molds, et cetera. That's a big financial commitment when the feds can come in and lock that facility down and take everything. But the moment the Prohibition ends, these become investible companies overnight.

And of course, if the Prohibition ends at a federal level, the market will go global even quicker. They say California cannabis is hands-down the best in the world. Imagine what the demand would look like for California cannabis if it could be shipped to every corner of the globe.

Stansberry Research: Prices would go through the roof.

Esposito: Exactly, but what people need to understand is that all this progress is self-reinforcing. While you don't see it in front of your eyes, progress begets progress. And that may be the most important catalyst of all.

Let me use insurance as an example because that's the area I know best...

I was speaking onstage at the CCIA's policy conference in February 2017. Everyone was asking me the same question: "How do we get more insurance carriers involved and other financial institutions?"

My answer was three simple words: "Buy more insurance." You buy more insurance, the marketplace will swell, we'll get more carriers, we'll get better rates, we'll get better coverages and terms and financially stable carriers.

Stansberry Research: What do you see as the biggest dangers investors in pot companies should consider if they're investing privately in, say, a state-sanctioned medical cannabis dispensary? Or if they're buying over-the-counter stocks in these pot companies? What do you think the biggest dangers they should consider are, and what's the biggest mistake you think these guys are making right now?

Esposito: People are just throwing money at this industry and jumping into virtually any investment. You have to look at the people involved. One of my clients, Bloom Farms, makes a vaporizing pen out of Oakland. They call it their "highlighter." This company is a rising star in the space. It's meticulous in every regard. But if you slow down and look at its team of executives, they have strong, and accomplished pedigrees. They have a winning culture.

Most people aren't looking at this like any other investment. They are mesmerized by the hype of marijuana. They should be asking: Who's the founder? What kind of management team do they have in place? What is their business plan? What makes them unique? How likely do you think it is that they're going to succeed?

If they're great problem solvers, if they manage the fundamentals of business, if they're effective in hiring and keeping talent, and if they're innovative... they're going to succeed in the cannabis space just like they would in a more standard space.

So really looking at that management team is critical. Sure, one guy can grow flower or make a great cookie or whatever it might be. But he or she is just an individual person. You need to ask yourself what the team looks like. Are they duplicatable?

That's why I love Bloom Farms. It has a charismatic leader, Mike Ray, and everybody I meet over there is just incredible. I say, "Whew, wish I could get in on some of that stock!"

Stansberry Research: I assume this company is still private?

Esposito: Unfortunately for both of us, it is still private.

But there's another critical thing to look for...

You see, underwriting insurance for a company is a lot like stock analysis. And for me, the second thing I always try to keep in mind is if the company has focus. No one can be everything to everybody.

What niche do you want to carve out? What do you want your client to look like? What kind of experience do you want your product to produce?

The answers might be things you'd never expect...

I had one guy tell me – this is interesting – he started a vape pen company because almost none of them are made in the U.S. The brilliant part is that U.S. dispensaries are beginning to realize that selling cheap vape pens of poor quality is exposing them to manufacturing and reputational risk.

If there was something wrong with a vape pen made in China or Taiwan or elsewhere and a customer sued, the liability falls on the U.S. dispensary owner not the foreign manufacturer – or more specifically, the first place of origin that the product was taken into the U.S.

If the product isn't made in the U.S. or the manufacturer doesn't have a U.S. domicile location where the pen gets imported to, there's no place for lawsuits to attach to from a technical standpoint. After all, nobody is going to go to China or Taiwan and be successful in a lawsuit. So this guy has a nice exploding niche.

He's going to be ready when Anheuser-Busch comes in and throws hundreds of millions or billions of dollars at the space.

Stansberry Research: So you see a future where there will be mass consolidation throughout the industry?

Esposito: I definitely think so. It's a natural part of an industry's maturing process. And it's going be like everything. Let's use the beer analogy...

Here's your Coors and your Coors Light. My son is in college, and I'm told that is a brand that is very inexpensive and students consume a lot. But you're also going to have your Lagunitas Brewing Company, your Knee Deep Brewing Company, or whatever your high-end beers are. Or wine for that matter – a boxed wine all the way up to your Screaming Eagle and Opus One.

It's the same thing with cannabis. It has to be a consistent product that delivers quality every time. If I find a product that doesn't give me the munchies, doesn't give me a headache, and doesn't get me too groggy... I will want to experience that again. If I can get that time and time again, I'm going to buy it 'til the cows come home.

– Chapter 5 –

This Man Set the Foundation for the Cannabis Economy

A conversation with Aaron Salz
Founder and CEO, Stoic Advisory

November 15, 2017

The following four interviews come from a business we greatly admire: RealVision.com, the world's first video-on-demand service for finance. Anyone with a membership can get instant access to the thoughts and ideas of some of the smartest investors on the planet.

RealVision.com has created a treasure trove of video content covering all topics of finance and investing, from ground-breaking documentaries to insights from the biggest names in the industry.

Few other outfits have studied as closely the issues and opportunities investors need to know about the marijuana industry before it's too late to benefit from them.

The four RealVision.com conversations we feature here describe the birth of a new industry in which millionaires and billionaires will be made... 54-year-old women with Gucci handbags are the demographic of the future... picks and shovels are the place to be... and the opportunity awaiting those who invest on the ground floor of the cannabis space is incredible.

The first interview from RealVision.com is with the first equity analyst in cannabis in Canada, Aaron Salz. Aaron pioneered research coverage of the medical cannabis sector, helping set the foundation of the cannabis economy.

His company, Toronto-based Stoic Advisory, offers financial consulting services to companies within the cannabis industry, both in Canada and around the globe.

Prior to founding Stoic, Aaron served as an analyst for medical and recreational cannabis companies at Dundee Capital Markets (now Eight Capital). He was also an investment analyst at specialist investment firm Interward Asset Management. While there, he played an active role in capital allocation, including investments in the cannabis sector.

Aaron holds a CFA designation and obtained his Honors Business Administration (HBA) degree with a focus on finance from the Ivey School of Business at the University of Western Ontario. He has been frequently quoted in the press, including *The Financial Post*, Bloomberg, Reuters, and *The Financial Times*. He is also a regular speaker at industry events.

In this interview, Aaron talks about...

- Why Canada is the best market for investors interested in the cannabis space.
- Pitfalls of the cannabis industry.
- And which companies – medical vs. adult-use recreational marijuana – will perform the best for investors.

RealVision: Aaron, tell me how big the opportunity is in the marijuana space.

Aaron Salz: Yeah, so I think if you look at Canada alone, we're talking about a C$20 billion-plus opportunity. And then if you look at the U.S., $50 billion-plus opportunity. And then globally, it just compounds from there.

So when you think about the global cannabis industry being a $100 billion, $200 billion-plus opportunity, there's certainly the ability for entrepreneurs and new business owners to really mint millionaires in this industry but also probably billionaires in this industry.

And I think we're just in the first inning of seeing who those business owners are going to be. But you can be certain that we'll have millionaires, and we will definitely have billionaires.

RealVision: Put that in context for me. How big is cannabis compared to other industries?

Salz: So if you look at the global alcohol industry or even just in Canada alone... If we think of the Canadian market opportunity as being close to C$20 billion for cannabis, that gets close to rivaling alcohol and beer specifically. So when you compare it to alcohol or more regulated and mature industries like that, you can consider cannabis one day getting close to competing certainly with alcohol. And you can even, one day, see it getting closer to even tobacco.

RealVision: Aaron, is Canada the best market for investors looking at the cannabis space?

Salz: I think Canada certainly is the best market today for investors. You contrast that to the U.S. – which is definitely getting more mindshare in the media – to Canada... Here we have a legal and national framework for medical cannabis. And we're getting very soon into the first G20 country that's getting into a national, legal framework for recreational cannabis.

So I think from an investment standpoint, Canada is definitely the most exciting. And also, really quietly, Canada has become the No. 1 legal exporter of medical cannabis in the world.

So I think for the medical cannabis opportunity and the recreational cannabis opportunity, Canada, by far, is at the forefront, which I think for investors makes it the most exciting opportunity.

RealVision: So tell me more about that. Canada today exports marijuana to other countries in the world?

Salz: Yeah. If you look at how medical regimes evolve globally, they always start in a very similar fashion. They start with very strict medical programs with a very limited set of qualifying conditions. That is, you can only be prescribed medical cannabis if you're a cancer patient or if you're a child with epilepsy or some sort of severe conditions.

And usually when markets start that way – like Germany, Australia, Brazil – those countries instead of setting up domestic production will import product from other countries. Canada is widely now looked at as the country with the strictest regulatory regime. And therefore, from a quality product standpoint, it's amongst the most trusted.

RealVision: Germany is a fascinating example, right? Here is a country that has 70 million people and had a system where they went out for bid. They said you can bid to grow cannabis and import cannabis into our country and any country in the world could bid. And only two countries qualified?

Salz: Yeah, so the German program was interesting. I think it sets precedent for what you'll see happen around the world. And that is that countries that are now trying to get into medical marijuana, or at least legalize medical cannabis then eventually probably pass recreational or adult-use cannabis, they're looking to countries that have a proven and well-thought-out program. And so, Canada becomes that obvious choice.

Germany is a perfect example where there was a tender process to get a medical domestic production license. And the only way you could actually submit your application for that is you had to have a proven history of compliant production. And so, when you look around the world, where is there really a proven history of compliant legal medical cannabis production at a national level? Canada and the Netherlands are basically the only countries that qualified in that set.

RealVision: As a former analyst in the space, how do you look at investing in the public cannabis markets?

Salz: I think when you look at all the public cannabis companies now, there's a few different opportunities available. On the major exchanges, like the TSX and the TSX Venture, you have over 23 federally licensed producers like Canopy and Aphria – companies that have existed for many, many years now and have attracted hundreds of millions of dollars of institutional capital.

Then you can go down to the CSE, which would be considered somewhat of a lower exchange, and in many cases on that exchange you have opportunities to invest in late-stage applicants that have yet to actually receive their license. In those cases, those companies are smaller than what's listed up on the TSX. You have companies that are trying to pursue R&D strategies. You have companies that are trying to pursue tech solutions that will be ancillary to the industry. And you also have the opportunity on the CSE, which is unique to the CSE, to invest in the U.S.-Canada sector.

RealVision: So you could invest in the Canadian Stock Exchange in the U.S. cannabis sector?

Salz: Yeah, so what's unique to Canada is not only are we developing truly industry leading and world-leading IP around cannabis, but we've also created the No. 1 capital market available for cannabis companies.

So really, I think Toronto and Canada are becoming the Silicon Valley of cannabis for the world, where companies globally – from Israel, Germany, Brazil, and of course the U.S. – are now coming to Canada to access capital because the markets here are just so much more open.

RealVision: Right. The largest grow in Uruguay, public in Canada.

Salz: Yeah, exactly. I think a lot of that has to do with the regulations here. Again, we have a legal and national framework that supports our regulated cannabis industry. And a lot of the companies here are banked by the big banks that many people are familiar with – RBC, CIBC, TD, et cetera. You go down to the U.S., and it is still a Schedule I federally illegal drug. As a result, the major exchanges have yet to actually list cannabis companies.

So companies from around the world are coming here to the Canadian capital markets to get funded, list their companies, and pursue strategies globally, not just in Canada.

RealVision: Wow. So back to investing in the space. We have the big companies – the Canopys and the Forias. We have the smaller craft growers – companies like Doja. I'm curious more about what pitfalls are there in this space? If you were looking in investing, what should you avoid?

Salz: Something that companies or investors should really be mindful of is investing in late-stage applicants. There are over 2,000 companies that have applied today, less than 70 have actually been licensed. If you [are told] that a company is a late-stage applicant, you have no clue whether or not it's next in line or 2,000th.

You have to be really mindful investing in companies that are advertising themselves as late-stage applicants because you just don't know how close they truly are. And you could be waiting for years.

RealVision: Aaron, what's more interesting as an investor – the medical space or the adult-user recreational space?

Salz: I think the opportunity in front of us today is definitely the adult-use space. It's coming in Canada around [July 2018]. And I think everyone is really focused on that. Again, you have an industry that's kind of in the shadows right now that's converting to a legal channel. So I think over the next few years, investing in the recreational market in Canada, then with opportunities in the U.S., is probably the most exciting because that's an explosive market that's just about to happen before our eyes. And we almost know it's there.

When you think of the medical cannabis market, we are trying to uncover and create opportunities that aren't necessarily there yet. So companies are investing in R&D and IP trying to replace pharmaceuticals like opiates. And we don't necessarily know that's a guaranteed path to success.

We're almost quite certain that people smoke cannabis, and they smoke it for pleasure. And when that goes into the legal channels, companies are going to likely do quite well.

This interview was edited from its original form to help with comprehension.

To watch the original interview, visit www.RealVision.com.

– Chapter 6 –

Canada's Most Connected Venture Capitalist

A conversation with Matt Shalhoub
Managing Director, Green Acre Capital

November 15, 2017

In our second RealVision.com interview, we hear from Matt Shalhoub.

Matt is the managing director of Green Acre Capital ("GAC"), Canada's first cannabis-focused venture capital fund.

GAC has made investments in many different areas of the cannabis space, including loyalty marketing technology for the cannabis industry, cannabis biotechnology and testing, cultivators, cannabis lifestyle brands, and much more.

Prior to GAC, Matt worked as the director of operations and business development at National Home Services, where he oversaw an annual capital expenditure budget of C$30 million and successfully led a C$35 million acquisition.

Most recently, he was an investor and general manager of ORCA, a venture-capital-backed clean technology business in the food waste-disposal industry. He helped lead the company from zero to more than C$4 million in revenue.

Matt was also an early investor in Aphria – a licensed producer of cannabis – which grew from a C$200 million company to a C$2 billion company in three years.

He graduated early with honors from the University of Western Ontario, receiving an HBA degree from the Richard Ivey School of Business.

In the following interview, Matt talks about...

- Canada as the gateway to the rest of the world.
- Where the next wave of success in the cannabis industry will be... and how investors can orient themselves for it.
- Which businesses he believes will become multibillion-dollar companies in the cannabis space.
- And the top three ancillary companies to keep your eye on.

RealVision: Matt, your fund, Green Acre – the first cannabis-focused venture capital fund in Canada – has made investments in many different parts of the cannabis space. Tell me more about where Green Acre has placed money.

Matt Shalhoub: We launched the fund at the beginning of 2017. We started making investments around March. Of the investments we made so far, we're invested in brands, software, hardware, and accessories. Looking to make a few more investments in some of those verticals. And looking at others like labs and some kind of product development R&D.

We're really most focused on non-cultivation opportunities at this point because we think it's part of the market that's really being under-capitalized and under-served right now.

RealVision: Interesting. So non-cultivation. If the first wave of cannabis investments was growing, you think the second wave is going to be an ancillary businesses?

Shalhoub: Definitely. Before Green Acre launched, we had been early investors in some of the cultivators. And I think as this market really started getting going, that was the only place to deploy capital. And obviously, the market really ran.

Now, as you look at this market and you compare it to other legal markets, it's almost inevitable to see a lot of companies pop up that are not just growing. You'll see companies that are taking that and value-added processing it to make edibles, or filled vape pens, or much more sophisticated forms of consumption – metered dosing, extended release tablets, transdermal patches, creams...

As this market evolves, what will happen here is what's happened in most of those markets where you've seen dried cannabis become an ingredient. It gets commoditized, and that raw ingredient now becomes very powerful. And the companies that can use that ingredient in other types of products become much more valuable.

The labs that are testing it will increase in size infinitely, along with the broader market. I think the software providers with very industry specific software can grow exponentially.

But those are a lot of the companies that aren't getting a lot of investor attention today. And I think, as a result, represent more attractive opportunities.

RealVision: Can you tell me a bit more about, in the ancillary market then, whether Canada or the U.S. is the best place to put your money?

Shalhoub: It very much depends on the sector within the ancillary market. What I mean by that is, the edibles market in the States for example, is a very challenging operating environment for an edible company because you need to have a facility in every single state that you want to operate within. So it's hard to see the scale that you'd see in a market like Canada where you can produce for the whole country out of one facility.

So markets like that – to duplicate all the equipment, all of the facilities required, to have all the licensing in place to be able to produce your product, different packaging and labeling requirements state by state – it's very difficult for those companies to really achieve the size and scale that companies here have been able to and will continue to be able to achieve.

So that's the market, for example, that I think is a very difficult market to operate in. Now if you're a cannabis software company, for example, the market's multiples of the size in the States, and it is scalable. Typically, the same or similar software can be used in most states. So I'd say if you're a cannabis software company, for example, the U.S. is potentially a more attractive market based on the size of that market. It really depends on the vertical.

RealVision: And I've heard Canada might be the best gateway to the rest of the world. Can you tell me a bit more about that?

Shalhoub: Yeah, that's very much our belief as well. Canadian companies in this space will become very long-term global leaders in this industry, which really attracted us to focus on the Canadian market as well. I think being able to operate in a market with federal legality is a much bigger benefit than I think a lot of people give credit to.

I really think Canada is ahead of the curve being able to bank, being able to raise real amounts of capital, being able to ship across the whole country, and being able to export to other federally legal countries. Canadian players will be very long-term dominant in this space because of that.

RealVision: So if I'm a retail investor and I'm thinking about placing my money in the space, what should I do? Where should I place my money in the cannabis industry?

Shalhoub: It's tough right now as the retail investor because the majority of the publicly available options to invest in today are mainly cultivators. With that lens on, I do think that there's a handful of great cultivators up here that make great investments.

As a retail investor, I'd be even more focused on continuing to follow the space very closely as it evolves to be able to make investments into larger ancillary businesses that become public and allow, obviously, the retail investors to buy shares because I think that's where the next wave of success will be in the space.

RealVision: Tell me a bit more about some of the red flags. Where should people not invest in the cannabis space?

Shalhoub: It's a bit of a tough one. I could speak to some of the areas that we've been shying away from. And I think the reason why we've shied away could be different than the reason some investors are looking. But we've shied away from U.S. plant-touching businesses.

RealVision: What do you mean by "plant-touching"?

Shalhoub: Anyone growing it, anybody retailing it, anybody extracting it. From our perspective, I firmly believe the U.S. will federally legalize in the future. But I think until it does, those will be harder investments to monetize, harder investments to be successful. There's tax-code issues. There's banking issues. So we have personally chosen to shy away from those types of investments. Within the next couple years, those could become very attractive investments, though.

As this market evolves in the States and Canada and globally, you're going to have to keep shifting your focus to other parts of the space. And that's the choice that we've made so far. I very much expect that to evolve over time.

RealVision: How big do we think the ancillary market could possibly be? If Canopy is C$5 billion, is there going to be an ancillary business that's also C$5 billion?

Shalhoub: I very much believe that there will be. [Accounting and consulting firm] Deloitte had put together research in 2016 that the size of the Canadian domestic market is over C$22 billion if you include the ancillary products and services. And I'd say most people size the actual cannabis market – meaning dry cannabis, oils, edibles to maybe C$10 billion in Canada. So with that math, you're thinking there's still probably C$12 billion or so of ancillary product and service business in Canada.

So I think there's a ton more market that will open up. I think that for sure there will be a multibillion-dollar Canada software company. There will be a multibillion-dollar cannabis hardware company. And what we're looking to do at Green Acre is to get in early to those businesses and provide them capital they need today. And provide them more than just that capital, but help to grow their businesses and turn them into those much larger, more successful companies.

RealVision: Can you tell me about the top three ancillary companies, then, to keep your eyes on?

Shalhoub: Yeah, I'd say a lot of them are, unfortunately, companies that people probably haven't heard much about because these are private companies. But we're obviously fans of Tokyo Smoke. It was our first investment in the fund. I think what Tokyo Smoke and others like that are doing with branding will become much more valuable. Starting to see big alcohol enter the space in Canada just shows you that these large brands see the value in the space, how it could disrupt their existing industry. So companies like that will be well situated.

Another portfolio investment of ours is a company called Ample Organics doing seed-to-sale tracking software. People haven't given enough thought to the customized software required in this space. And I do think that, to my point, there will be a billion-dollar-plus cannabis tech company in this space.

RealVision: Right. Ample powers what percentage of the legal cannabis growers?

Shalhoub: About 2/3 of the producers in Canada are on its platform already. But most haven't heard of the company, and that's my point. A lot of the investing community out there has yet to even hear the names of a lot of the companies that I think will be very large players in the space.

And the third, I would say, it's harder... it's even unnamed. But I do firmly believe there will be a hardware and accessory business making vaporizers, or grinders, or all those ancillary products that will be a billion-dollar-plus company. I think potentially even created by the acquisition or merger of a number of those companies in the space.

It's already a multibillion-dollar market in North America, but it's very fragmented. That's one area that'll certainly get brought together and form a much larger company.

RealVision: What are your thoughts on the activists? The Jodie Emerys of the space?

Shalhoub: The activists are a large reason why the industry is where it is today. I think the medical market wouldn't have happened in the early 2000s had it not been for activists pushing it through court challenges. So for that, I'd say we're very appreciative of all the work and time they put into this.

But I think a lot of those activists that are still in the space today are at a very integral point where they can make a decision to either try and flip over and join the legal market or they can continue to try and operate in the illegal market.

And I think those that are smart enough will make that transition over and could potentially make a bunch of money and be very successful in the space. I think those that continue to operate in the gray or black side of the market will be penalized. I think people that expect to continue to operate dispensaries in Ontario after it's legalized have lost their minds. The government has been very clear in its position here.

People will often say, "Well, the product's legal, I'm not going to get arrested for selling it." But cigarettes are legal, and people do jail time for selling illegal cigarettes. They're selling them outside of legal channels.

So I think that the government is going to throw the book at some people to make examples to get the illegal operators out of the market. But those who make that transition can do great, and we hope that they do.

RealVision: Is now the right time to invest?

Shalhoub: In certain markets it is. In the Canadian market, it definitely is. I think we're in the maybe second or third inning of a nine-plus inning game. So it's still somewhat early, but I think the timing right now is quite good to get into the market and benefit from what will happen in Canada.

RealVision: Thanks, Matt, for your time. It's been fascinating to hear about your perspective on this space. And it's amazing to get better insight into what will be the second wave of successes in the cannabis space.

Shalhoub: Thanks. It's always fun to talk about cannabis.

This interview was edited from its original form to help with comprehension.

To watch the original interview, visit www.RealVision.com.

– Chapter 7 –

The CEO of the World's Largest Legal Cannabis Company

A conversation with Bruce Linton
Chairman and CEO, Canopy Growth Corporation

November 15, 2017

In our third interview from RealVision.com, we meet Bruce Linton – founder of Canopy Growth Corporation (CGC) and co-founder of Tweed Marijuana.

His primary focus is to position cannabis brands in a competitive market and to raise the capital necessary to fund such operations...

Canopy was the first federally regulated, publicly traded cannabis producer in North America. In 2018, it's the largest legal producer in the world, valued at more than C$7 billion. It is the first company in the industry to be offered debt instruments by a federal lender. And its C$191 million investment from global beverage behemoth Constellation Brands marks the first major investment in cannabis made by a publicly traded company to date.

Bruce has been responsible for the acquisition and/or disposition of more than C$600 million in business assets – including three merger-and-acquisition transactions valued at more than C$500 million total since founding Canopy.

He has established regular engagement with the World Bank and Asia Development Bank while focused on markets in India, the

Philippines, China, Peru, Colombia, Ecuador, Azerbaijan, and Uzbekistan.

Before launching Canopy, Bruce served as the chairman of the Ottawa Community Loan Foundation, board member and Treasurer of Canada World Youth, on the board of governors for Carleton University, president of the Nepean Skating Club, and president of the Carleton University Students Association.

In the following interview, Bruce discusses...

- Why he expects his company's revenue to grow dramatically in the second half of 2018.

- His target audience for recreational cannabis: a 54-year-old woman with a Gucci bag.

- And the Internet revolution versus the cannabis revolution.

RealVision: Bruce, we're sitting here in the lobby of Canopy... this incredibly beautiful lobby of this really established business. How did you get started in this space?

Bruce Linton: Well, I like the space. It's the opposite of a mullet. It's kind of like the party in the front, and when you go back there, it's all business. And what got me into this space is, it was pretty clear two governments ago... They were kind of like the Tea Party, right?

RealVision: What year was this?

Linton: This was the conservatives. It was about 2012. And there was a lot of discomfort with the prior regulation from a policing governance perspective. But it was obvious that there were a lot of people who were accessing cannabis through it. And what was going to happen was there was going to be a complete turn in the supply chain. It was going to become a governed product, which meant narcotic.

When I looked at it as a tech background person, I thought the regulators are really going to care quite a lot that you don't lose the product... that you have a total chain-of-custody logic.

And so when I approached the sector, I approached it from the perspective of how will I make those who are going to govern me comfortable that we have a really well understood method of not losing the product? And then we will build brands on top of that because they're going to regulate us and give us licenses because we'll make them not be embarrassed.

What attracted me was it was pretty evident there were a great number of people who liked cannabis. A change in the supply chain for something for which there was a huge market has never happened in my life and won't happen again – where the government inserts itself to create a market for something that's ungoverned and black market and one of the only big commerce cycles in our whole world that the government has none of.

All of that made me go around to about four or five people. And the first four when asked, "Would you like to start a marijuana business with me?" Went, "No, that is a very bad idea." And one of them offered that it will probably be the worst idea that you have, and maybe you're going to get killed by the bike gangs.

That confirmed to me that it was a very good idea because all the people who I approached who I thought were business people and rational rejected the opportunity.

This place [we're sitting in] now is the former Hershey chocolate factory. It's about 450,000 square feet. It has room for about another 250,000 square feet. And when I started, we occupied the first 180,000 square feet.

What I liked about it is, if you think about being governed by Health Canada, and if you think that you're going to land in a class of products that are going to be on par with narcotics... when you say the words "narcotic" and "government," that means C$2 million a year of overhead, in the first years, to make sure you're complying [with] security, QA, recordings, all of the things that you would expect.

If you have a very small facility, you would never make money. But I can tell you when I started with a really large facility, the view was, "You are quite stupid because there is no market for that much cannabis." And I kept thinking that's really kind of short term.

We are now tripling the amount of space we use. We occupy with our growth area about 168,000 square feet. And so, when you look at our infrastructure here, we've gone from making air conditioning in January to having heat exchangers that cut my consumption of electricity to 50% as soon as it's minus 2 degrees centigrade or colder.

We've gone from single story growing ops to double-decker. And when you do double-decker, your cost per square foot goes to half. Your infrastructure is used better. And so, what we've been doing is making better product in a better operating environment, which means that the cost is less, which means that we can be extremely cost competitive on the base product.

But in reality, the base product is almost only an ingredient. You come back and sit in this chair with me in three years, and we're going to be talking about the massive array of products which we make which use that ingredient for medical and adult enjoyment. And the reason that's going to be the case is people only buy dried cannabis now and roll a joint because they haven't had as many good options.

What's going to happen is the options are going to be formed and the intellectual property is going to be around them. People will walk into venues or go to doctors, and they're going to get branded products that achieve an outcome. That's it.

RealVision: I'm fully sold on that. I often think of hops. I like drinking beer, and I don't know anything about hops. I don't know where hops are grown. I don't know what makes good hops. I just know that hops are grown efficiently somewhere in a thoughtful process and then end up in my delicious beer.

Linton: Yeah. And you know there's hops in it because they told you, but you didn't really know there was hops in it.

RealVision: No, I know there's hops in it because I'm in the cannabis industry and they're similar to cannabis.

Linton: We just did a deal recently – a small one, may not have noticed it – with Constellation Brands.

RealVision: Tell me more about that deal.

Linton: Constellation Brands is a multi-branded international leader in terms of beer, wine, and spirits. For the most part, if you're going to make beer, you probably don't necessarily own the hops. And if you're going to make vodka, you probably know whether or not it has certain ingredients in it, but you don't control and own them.

RealVision: If you make coffee... Starbucks doesn't own its coffee plants.

Linton: Right. And so, you can see the migratory path that's going to occur. And there will be people who are very good at being subordinated. The current system – and many countries will keep it the same – is you do need to grow the product so you can vertically integrate because if they go to a completely open supply chain, the black market is back in business. And it's called the supply chain up.

I think you'll see a lot of rigor, which will keep the prices not falling at a standard commodity curve. But the value proposition is much higher when you change the format.

RealVision: So with Canopy, this is just one small piece of the Canopy empire. Tell me more about the scope and scale of your business.

Linton: The empire, um, we're really working on a Death Star. It's going to look great. We currently operate in six provinces and six countries.

RealVision: What countries?

Linton: First and foremost, Canada because Canada figured out public policy. And a lot of people will say that Mr. Trudeau is going to legalize cannabis in July 2018. It's actually not true. Cannabis access was legalized in 2001.

RealVision: A very long time ago.

Linton: What he's going to do is expand the access. And that's really important because public policy cannot occur in a void.

We've had a very focused interaction with cannabis physicians and regulation since 2001. The reason I highlight that is no other country in the world has been so active on that, which has resulted in fantastic public policy put out and evolved.

Which means by being really good here, then you can think about Germany, Denmark, Brazil, Australia, Spain, Colombia, Chile. And so, the list of countries I mentioned are federally lawful, taking a process which is governing the production and distribution. And those are domestic markets which have a real chance of being a real market.

They're not people running around saying, "Well, we can grow it for nothing, and we're going to sell it in the whole world." Because this is still a United Nations-restricted narcotic. It is not getting on buses and going everywhere.

The interesting thing is Canopy is the first company – and I still think only – making good manufacturing practices (GMP)-certified gel-caps, which are an identical dose format every time.

That makes doctors a lot more comfortable with this as an option, which means the number of patients will probably double [from the current 200,000 Canadians who have medical rights]. And it means that what happens in the back half of 2018 is we're going to have three to four million people who want to try it, which means they've got to buy it.

And so, I suspect even our rapid growth might be overtaken by what could be even more rapid revenue.

RealVision: In the world of the Internet, you had to convince people to use the Internet. In the world of cannabis, people already use cannabis.

Linton: Well, there are [people who already use cannabis]. But I'm targeting the 54-year-old woman with a Gucci bag who has not bought any at least in the last 25 years. And what I want to make that person comfortable with is that this is in fact a very thoughtful alternative to buying white wine.

And the reason is, if I can make something which has no or low calories, moderate to no drug-on-drug interaction, that triggers more of a giddy, euphoric feeling than because you're having a white wine... you might have a bit of fish or something. I think that demographic is really the right demographic for me.

I would say there is a bit more of a parallel to convincing people to use the Internet and convincing that this might be a suitable alternate to the other things they would have otherwise purchased.

RealVision: So do you think the future then is the adult-use market? The future then is getting people to swap out current psychoactives or substances in their lives and have them move into cannabis?

Linton: There's that, and I think in medicine. In fact, the more I think about it, the more I disagree with you – which is really dumb – because nobody had Prohibition against the Internet for 80 years. Nobody had an absolute, unfounded bias against the Internet.

So I'm actually fighting a far more uphill battle, I think, than Google did. I have to convince doctors. I have to do twice as good a job to convince banks. I have to do twice as good a job to convince patients they should swap out something. I have to do twice as good a job to convince people buying alcohol that this might be a suitable alternate.

The reason I like brands, and promises and fulfilling them, is there are so many negatives that we will overcome... the brand loyalty and the connectivity to a quality outcome will be far superior. Because you know what? Somebody said Yahoo will never lose. Yahoo's the search engine. And I don't know, I bet there's someone who still uses it, but I'm pretty sure Google's doing OK compared to them.

And so, I think that the transition time in the Internet is much more rapid than we'll find in the cannabis space. I suspect we're going to have a lifetime of loyalty if we change people's embedded negative perception.

This interview was edited from its original form to help with comprehension.

To watch the original interview, visit www.RealVision.com.

– Chapter 8 –

The Hidden Value in Legal Marijuana

A conversation with Todd Harrison
Chief Investment Officer, CB1 Capital Partners

November 15, 2017

In the fourth interview from RealVision.com, we hear from Todd Harrison.

Todd is the founding partner and chief investment officer of CB1 Capital – an investment manager specializing in medical cannabis solutions, products, and therapies.

He based his company around the idea that cannabis will disrupt Western medicine as we know it. His goal is to educate the next generation about the inevitability of medical breakthroughs with cannabis going mainstream across the globe.

Todd has spent almost three decades on Wall Street managing risk and researching market strategies. His institutional experience includes vice president at Morgan Stanley Global Equity Derivatives, managing director of derivatives at The Galleon Group, and president of hedge fund Cramer, Berkowitz. Todd is also the founder and CEO of Minyanville Media as well as the founder and president of The Ruby Peck Foundation for Children's Education.

His first book, *The Other Side of Wall Street*, was published in 2011. And he is a contributing author to "Threat, Intimidation, and Student Financial Market Knowledge: An Empirical Study," published in the *Journal of Education for Business*.

Todd has appeared on CNBC, FOX, FOX Business, MSNBC, and Bloomberg as a financial authority. And he has been quoted and covered in *The New York Times*, *The Wall Street Journal*, *BusinessWeek*, *Worth*, *Fortune*, *Forbes*, Barron's, Dow Jones MarketWatch, *New York Magazine*, and Canada's *National Post*.

He was also featured in the 20th Anniversary documentary of Oliver Stone's *Wall Street* and the 2013 documentary *Money for Nothing: Inside the Federal Reserve*.

In the following interview, Todd identifies...

- The biggest misconception about the cannabis industry.
- 10 areas he thinks will provide the best opportunity set for the cannabis space.
- Unexplored growth for medical marijuana.
- And unlocked uses for cannabinoids beyond CBD and THC.

RealVision: We're here with Todd Harrison of CB1 Capital Partners. CB1 is your latest venture, but you've got a long history in the hedge-fund space. And not only in hedge fund, but post-2001 you founded a company called Minyanville, which focused on public education. Could you give me a quick snapshot of your background in the lead up to CB1, and we'll talk about the interesting stuff you're doing there?

Todd Harrison: Sure. I graduated from Syracuse University, a proud Orangemen. I graduated in 1991. About three days after graduation, I found myself on the equity derivatives desk at Morgan Stanley armed with a Black-Scholes model and a *Wall Street Journal*, neither of which did me much good. But I taught myself the ropes. Was there for about seven years. I was promoted to vice president at the age of 26.

I went over to a shop called Galleon Group on the buy side where I managed their derivative portfolio, and I was there for two years.

Subsequently left and went to Cramer Berkowitz, which is a $400 million fund, and I ran derivatives and trading there. I was a partner and operated that fund through 2002.

Subsequently to joining Cramer Berkowitz I was working with Jim Cramer, obviously, and he had asked me to fill in for him one day writing TheStreet.com column. I did as a favor, and I took to it. I enjoyed it. I enjoyed the synthesis that it provided, and I enjoyed the notion of helping people, which is really why I started Minyanville after 9/11 with the idea that trading in capital markets wasn't really providing much societal redemption. After 9/11, I wanted to do something that was meaningful.

So I started the Ruby Peck Foundation for children's education where we help children. And I started Minyanville to effect positive change to financial understanding. We did that for about 15 years. Helped a lot of people avoid a lot of heartache with the financial crisis and such. And we like to think we empowered people to make better and more informed financial decisions.

It was during my time in Minyanville – about 2010, 2012 – that I really started to take an active interest in the cannabis space. But not for the focus that I'm presently on. Really, it was more pragmatic. Tax revenue, job growth, prison population, crime rate, these things which seemed intuitive to me. But obviously, there's a pretty pronounced campaign domestically regarding the status of cannabis.

Over the past three years, I've worked with, who is my now my partner, Warren DeFalco. He cured himself from onset epileptic disorder about 10 years ago and has subsequently read every piece of literature, every research note, every FDA trial that involves cannabinoids. We started to trade together and make some money. And lo and behold, after looking for about six, seven years to enter this space, I found what I perceive to be an advantageous risk/reward in approaching this space.

RealVision: Just to make the link there, CB1 is the name of your firm. CB1 is also the receptor tied to cannabinoids, correct?

Harrison: Yes.

RealVision: This is a physiological phenomenon. This is part of the human body.

Harrison: That's one of the core competencies of this thesis. Human beings – as well as cats, dogs, plants – we all have what's called an ECS, or an endocannabinoid system in our body. These are receptors throughout our body, throughout our extremities and our organs, that are natural receptors for what's called endocannabinoids. Endocannabinoids are produced by the human body.

What's happened over the last 50 years or so, is with the advent of fast food – and really fatty acids and all of the junk that we are putting into our bodies – the ECS, endocannabinoid system, has really gone through what I've heard described as a depression. So our endocannabinoid receivers have really been ignored for a long time.

What the compounds in cannabis, the cannabinoids, do is they target particular receptors in your body. So Western medicine, for instance, you take a pill and you hope it gets to where it's going. [The cannabis] approach, which we really do believe is disruptive health care, this is going to disrupt health care and how it provides efficacy for a wide range of unmet medical conditions. You can take CBD, for instance, proactively as a wellness supplement or nutrient, or you can take the compounds in a more concentrated form, and reactively address diseases which are currently unmet medical conditions across the Western medicine sphere.

RealVision: And so just to link that back, I mean the receptor, or the endocannabinoid system, is this a general-purpose receptor? I can use it to take anything across the body barriers to this type of treatment? I can attach something else to THC or to CBD?

Harrison: Well, it's a magical plant and it's also a mysterious plant. You have to remember that for the last 50 to 70 years, you were not allowed to research cannabis due to U.S. regulation. It was deemed the devil's weed and Reefer Madness… It's been a bit of a propaganda campaign on behalf of the U.S. government to really drill in that this is a gateway drug and that this is a very bad feeling.

For sure, if you're under 21, it's terrible for you. If you're under 21, each time you ingest cannabis is incremental damage to your developments. Under 21, it is something that should not be touched. But as you get older, interestingly, it becomes better for you.

There are studies now for dementia, for Alzheimer's, which makes sense. It stimulates activity in the brain for these older people where their brain activity is slipping. There's about 85 FDA trials currently in place that focus on the cannabinoid-based research, and those are the companies that we're really focusing on coming out of the gate.

RealVision: So the science is very clear that there are positive benefits associated with this therapy. Alcohol has similar dynamics. You shouldn't be drinking alcohol when you're under the age of 21. You shouldn't be ingesting many things when your body is developing that have a different role or a different characteristic as you age. And so, that's intuitively very clear and makes a lot of sense.

And yet, it has this interesting wrinkle. This is a little bit like investing in the alcohol industry post Prohibition. The difference is… there was an established alcohol-distribution system. While it was "illegal" under Prohibition, there were still methods of distribution. There were still methods to obtain it, many of which were "illegal." But when Prohibition was lifted on a nationwide basis, these could immediately be public and open companies.

You haven't had anything like that [with cannabis] because it's illegal at the federal level. So you interestingly have companies you can invest in, but most of them really can't do business in the United States, certainly on a nationwide base.

Harrison: So we have a few operating theses here. Number one is that there's going to be drugs from state dispensaries, but there's going to be medicine prescribed by doctors as covered by insurance. It's that latter bucket that we're focused on for several reasons…

Number one, as soon as medical efficacy is demonstrated, the DEA by law has to reclassify, at which point the institutions are going to be much more aggressive entering the space, in my opinion.

We're looking at the industrialization. We're looking at the complex building out. So we're not a cannabis fund, we're an emerging health care fund that's focusing on cannabinoid-based solutions and wellness techniques. This is a wellness fund. And we do believe it's impacting investing. We believe this is going to save and help a lot of people's lives and struggles with unmet medical conditions.

So we view the ecosystem through the lens of 10 verticals that we think are going to provide the best opportunity set for the space. And keeping in mind, we're an emerging health care fund. We're looking at wellness and disease cure as our primary points of entry right now in terms of our risk profile, and what we're looking at from an investment standpoint.

But really of the 10 verticals, there are two that are real populated right now and are tradable. One of which is the cultivation and dispensary side of the equation, the growers, the LPs (licensed producers), and they're up in Canada. The other side is the biopharma space, which we're more focused on because we believe that the input is going to be commoditized. It may not be as commoditized as hemp, which will trade in Chicago.

[W]e think hemp is going to be Farming 2.0, the Renaissance, for the American farmer. But right now, in terms of the opportunity set and what's populated, you really have cultivation, and dispensaries, and you have the biopharma side. And you have some hemp ways. I'll get into those. But we're really looking at the biopharma side because we're not looking at the input.

We're not looking at the plant in and of itself. We don't touch the plant. We don't move the plant across state lines. And as a matter of course, we let the listing protocols for any of these stocks serve as our first layer of due diligence. So we're not touching any private companies. Out of the 500 or so companies, we've circled about 50 that meet our criteria. Of those 50, we'll have anywhere from 15 to

20 in our portfolio at any one time. We're looking at a diversified approach.

So ultimately, what we're looking to capture here aside from the biopharma (which we believe is going to be powerful) and the cultivation side (which we don't want to touch because there's so much supply coming on continually), there are things like industrial hemp and farming, which we think are going to be real positive plays. It's already the fastest-growing industry, but also the largest cash crop in the world, while illegal in the United States. So we think that the industrial hemp and farming side of the equation is going to be great.

We think that vanity and cosmetics are going to be huge. There are studies right now involving creams that remove wrinkles. And I don't know what the world is like by you, but by me, that's going to be a pretty good selling product.

RealVision: I would guess so. I would think the THC would lead to just not seeing [the wrinkles] as opposed to actually making them disappear...

Harrison: That's the problem. Glad you brought that up because that's a pretty popular misnomer. When I talk about this strategy, which is a wellness strategy and a health care disruption play, the first question I often get is, "Did you bring any samples?" I read articles about the brain cancer trials that are coming through the pipe, and they (GW Pharma) can't really reject because people are still surviving.

They said that in February 2017, they still haven't released the endpoints so we're now in September and people are still surviving from this treatment for brain cancer. It's powerful stuff what's going on here. So that's where we want to focus our initial capital, is finding those companies.

RealVision: And that really is part of the opportunity in near view is that there is a negative connotation associated with a drug that has been used recreationally for psychokinetic or psychopharma purposes. But really, most of the research that's going on in this space it's not as a hallucinogen. It's as a treatment for the identified receptor in delivering medicine in a unique and novel way that is tied in to the human body.

Harrison: Well, yes. So 95% of all cannabinoids are non-psychoactive, and I think that's a pretty big misperception about the space. And again, we've been fed this line of thinking about cannabis for so long that it's a gateway drug. And yes, it is a gateway drug. It's a gateway drug off of opiates. It's a gateway drug away from heroin. It's a gateway drug in the right direction, not a gateway drug toward those things.

But as we look at the space, and we start to talk about the cannabinoid profile and what is going to come down the pipe, it's pretty powerful across a wide range of wellness applications. And that's what we really are focusing on now.

So we're talking to a sophisticated audience here. RealVision is a more astute crowd than I think CNBC might be.

RealVision: Right. We like to think so.

Harrison: I'm looking at this as a four-way arbitrage, and I think your audience would understand this. We look at this as an arbitrage against price, before the institutions get in. We look at this as an arbitrage against time, before policy reverts... and we believe it will.

We believe this is an arbitrage in perception. So popular perception, this is about getting high. We believe this is about getting well, and we think that capturing that disconnect between perception and reality is going to be powerful for anybody who sees what we're seeing.

And then finally, it's an arbitrage on liquidity. Right now, exposure to the space is really focused on private equity. We're running a hedge fund. We're running a health care fund, an emerging health care fund where we're providing valuation on a monthly basis. We're providing liquidity on a quarterly basis. So from a risk-profile standpoint, we think that we have a competitive advantage on what's out there right now.

RealVision: When you talk about providing liquidity in a space that is relatively young, relatively unformed, has a lot of change, I would have to expect that you're anticipating levels of volatility that are relatively high.

Harrison: Of course. Any period of price discovery, any process of price discovery – and that's what this is – is going to be volatile. But I'll tell anybody who will listen that Jeff Sessions and Donald Trump are the best thing that could ever possibly happen to CB1 Capital because they've kept a blanket on prices, on the whole psychology surrounding this.

Psychology is important here. Perception is reality. Once the efficacy is demonstrated, once people realize that this isn't just about smoking a vaporizer on a Saturday night, this is actually about killing brain cancer and reducing epileptic seizures in children by 50%... *Cutting them in half...* And people start to understand the efficacy across not just one ailment, but a whole host of unmet medical conditions... We believe that that's going to be the real arbitrage.

From the liberal who smokes on a Saturday night and thinks that they're being liberal to the cannabis conversation to the entire world that really is uneducated about the medical benefits and the wellness attributes of the plant.

RealVision: Is that the process you think that brings this sector back out of the cold effectively? That it becomes socially acceptable because of the medical efficacy? It's not an improving social environment in which people are more tolerant of other people's independent activities?

Harrison: Well, it's both. So 62% of the U.S. population approves recreational legalization. I think the number is closer to 92% in terms of the medical efficacy. So from a pure demographic standpoint, as we start to look forward to the next cycle of elections, it's a matter of time before it's legalized. This is going to happen more and more because the constituency wants this, so it's political suicide to go out and try to kill this right now.

So that's the demographic shift. But the path that we're very much focused on is the efficacy side... We believe the DEA is going to respond by reclassifying, whether they reclassify just CBD (which is possible but not likely) or reclassify the whole plant remains to be seen. But if they reclassify CBD only, then it's just a matter of time before the Glioma (brain cancer) trials come through. That's a 1-to-1 CBD-to-THC ratio.

So this is happening. It's happening now. The opportunity is coming up quick, and we just hope to be in a position to capture it, and be in a position to sell it to some of these bigger institutions as they get into the market.

RealVision: And this is really much more classically like what your background in hedge funds would have been, right? I mean, starting hedge funds in the 1990s, like they were the scrappy, get out in front, buy stuff before other people are going to do it. They weren't the established institutions that they are today. It was a much more aggressive environment. Does this space feel like the early days of hedge funds to you?

Harrison: Absolutely. Now in understanding the perception of hedge-fund managers in the world, I was in that game for a long time. And after 9/11, I wanted to do something meaningful.

Coincidentally, the hedge-fund community has evolved into 10,000 people standing in a circle shooting each other looking for the same stocks. I had no interest in entering that space again. I have no interest in trading the FANG stocks back and forth anymore. I'm interested in this wellness play. I'm interested in this health care disruption play.

Wrongly or rightly, I genuinely believe that this is going to help a tremendous amount of people throughout the world with more fulfilling lives, or with life at all, save a lot of lives. And until that's the popular perception, I think the opportunity to capture that chasm between risk/reward remains.

RealVision: Do these trade at a valuation discount to the rest of the market, or are they trading at a premium because of the growth potential?

Harrison: They're trading at a premium based on the growth potential. But again, going back to our earlier conversation, the reason the LPs are unattractive to us is because everybody wants to be in the cannabis space and have exposure.

To borrow a line from the movie *Dave*... people are so thirsty that in the absence of water, they're drinking the sand not because they like to drink sand, but because they don't know the difference between drinking the sand and drinking the water.

Right now, the LPs are the only game in town. Everybody who is investing is investing in those. We don't like those because they're crowded. We want to find these companies that are phase one, even pre-clinical in some cases, who have the management team, who have the patent portfolio, who have the cash on hand, and a number of other filters that we lay across our portfolio companies ranking them in each category before we arrive at our overall portfolio ranking.

RealVision: So would you be able to short the LPs?

Harrison: I'm more inclined, honestly, to short market ETFs against our portfolio. We want to stay long the cannabis or cannabinoid-related names. And we want to stay as close to the edge in terms of hedging the downside as possible, but we only hedge through defined risk strategies. We have no interest in trying to time the market per se.

RealVision: So you mentioned your tier of 10 verticals. You talked about dispensaries and you talked about bio pharmaceuticals. What are the other types of applications?

Harrison: Pet foods and supplements.

RealVision: And you mentioned that cats and dogs like to get high as well. I'm sorry. I'm sorry. They have a cannabinoid receptor. So this is just built into the mammalian system, so it's equally efficacious for your pets.

Harrison: I mean, plants have a phytocannabinoid system. All living things have these receptors throughout their bodies or throughout their organism, as the case may be. So yes, it works on dogs and cats also. I have a pretty neurotic dog. I give him CBD every night, and he seems to calm down.

RealVision: Interesting.

Harrison: So it's understanding. I think education is another vertical, education in media. Education's going to be paramount to this whole process. I think people are going to crave education and understanding of what's going on. It's a pretty fertile market because of the landscape and the lines we've been fed for so long.

But the nutraceutical side... We believe that CBD is going to be infused in foods almost as a cost of doing business. I've already had calls with some potato chip companies who are asking about the benefits of this. We think this is going to be a pretty significant theme going forward for the next five 10 years.

RealVision: My understanding of CBD is that it is the non-psychoactive component. THC is the psychoactive component.

Harrison: Right. That's two.

RealVision: There's lots of other drug names, and various other components that go into the cannabis complex, or in the cannabinoid complex. When you're thinking about using something like this, and you see it as a nutraceutical, what would the benefit be of a potato chip with CBD in it, for example?

Harrison: Well, CBD in and of itself promotes homeostasis. So if you remember from earth science back in the day, homeostasis is your body's wellness, and the ability of your body to talk to each other and interact with the organs, interact with each other.

CBD has been demonstrated to promote homeostasis in the body, just like our endocannabinoids are originally intended to promote homeostasis in the body.

So again, this is not a mature space, but for people who are interested in those types of things, there's a fair amount of research online. We are following the FDA as closely as we can. We want to draft off the FDA as these companies come to market.

RealVision: So when you think about something like the nutraceuticals, do you see an FDA-recommended dose type dynamic, or would people eat the potato chips with CBD because of the health effects similar to the way that they will take St. John's Wort to improve their [mood disorders]?

Harrison: Well, our sense is that's it's all going to go through the FDA eventually. Yes, there's going to be state dispensaries because they're making so much money. But if the government doesn't act, if the FDA doesn't own this in the States, you're going to own it.

There's way too much money at stake. Follow the money, as they say on Wall Street. There's so much money involved here that the FDA will figure out a way to own it, and regulate it, and provide standards. This is going to be a process that's going to take some time for this to take hold, but certainly that's the trajectory we think it's going to land on.

RealVision: Is that similar to tobacco or alcohol? I mean, is it ultimately the taxes that are going to drive the legal acceptance of this? Will it be the government deciding this can be sold, but there's a universal sales tax?

Harrison: A combination of the medical efficacy is going to remove the demonization, if you will. But ultimately, it's going to be the money that the government, I think, is going to keep an eye on. And I'm not talking about just pure tax revenue. I'm talking about the money saved from overcrowded prison populations and the safety profile of a lower crime rate. These are all parts of the conversation that need to be factored in. I think as soon as the FDA demonstrates efficacy, then the conversation takes it to the next level.

RealVision: You know, it's an interesting societal question. Because I completely agree with you that the costs of enforcement, as we found with Prohibition, are remarkably high. It separates significant segments of the population from others. Creates criminal records where none should exist.

But if I think about the acceptance of alcohol, there was no need for the efficacy component. We're told a glass of wine at night is good for you. Whether that's true or not, I don't know the science well enough to articulate. But it's certainly not at the same level of efficacy of what you're referring to within the cannabinoid complex.

Harrison: Correct. And so I use the analogy, for 10,000 years, people have been ingesting cannabis. It's been like throwing spaghetti against the wall. They realize that there's a different psychoactive profile coming out, but they can't quite quantify it or qualify when exactly that is. The science is now catching up to the plant to the point where we can now take the strands of spaghetti off the wall and plug them into particular ailments in your body.

You talked about CBD and THC... there are a host of other cannabinoids that they are just now discovering but some of them are very powerful. CBDV is for adult epilepsy. THCV is psychoactive, but it reduces your appetite. So there's a lot of magic left in this plant that has yet to be unlocked.

RealVision: That's fascinating. When you think about how long this investment opportunity will exist, do you have any perspective on it? When are we going to find our first cannabinoid company in an index? Or, are they in indexes already? Obviously, there's an ETF side to any theme out there, but is there anything that is currently investable from a large institutional space?

Harrison: GW Pharmaceuticals is certainly leap years ahead of the competition. They're running the Glioma Phase 2 trials. They're going to file for an NDA [non-disclosure agreement] with the FDA in October. So we're hoping that that's the straw that breaks the camel's back, that medical efficacy. But again, if they turn around and say we're going to classify CBD, then it's a matter of time before the Glioma trials demonstrate efficacy for that as well.

We're looking right now at the companies where they're trading in the local markets. We see a handful of Israeli companies that we like, so we'll buy them on the Israeli exchange where the volume is. Same thing with the handful of Canadian stocks we like and Australian stocks that we like. The longer the U.S. takes to get on board, we view it as being incrementally positive for the companies on our portfolio.

RealVision: Where would you recommend viewers go to learn more?

Harrison: CB1Cap.com. This is not a solicitation as much as communication, but we're excited. To take a step back, I'm a passionate person. So I think anybody that is trying to be successful in life really must follow their passion.

Passionate about giving back after 9/11, I started Minyanville to help people make better and more informed decisions. I thought that was important. The media model ultimately shifted of course. But I see a similar calling, if you will, in terms of this space based purely on the wellness attributes and based on just the efficacy that it's going to provide across a host of currently unmet conditions.

RealVision: You mentioned that this is a disruptor model. By that, I'm assuming you mean that this will take the place of, or has the potential to take the place of, so many of the existing pharmaceutical products that people use. Does that suggest that this is a net opportunity for the pharma space, or is this ultimately yet another deflationary pulse that is going after high-priced branded or even generic products?

Harrison: I think it's a bit of both. I think Big Pharma – who's been the biggest lobby on the other side of legalization along with the spirits industry – views this as buy build. I think they're certainly on the hunt for these companies, but there's only a handful of them that are really viable right now. We certainly think it will be a buy build, but estimates right now are that the candidates will shave about 5 billion off the top line for Big Pharma when it comes legal. So there are certainly a lot of dollars in place that are trying to avoid that.

RealVision: If I'm thinking about the wellness aspect of this, are there other products, other chemicals, similar to cannabinoids that you see representing a portion of your portfolio?

Harrison: Opioids are similar from a profile standpoint. But the reason that you will die from opioid overdose but not cannabis overdose is because the opioid receptors are in your brain stem which controls your lungs. So if you ingest too much opioid, you're eventually going to...

RealVision: Stop breathing.

Harrison: ... stop breathing. Cannabis doesn't have that profile. There are no ECS receptors in your brain stem. So from an adverse effect standpoint, the adverse effect profile is much more powerful in terms of non-damaging than a lot of these competitors or current options in this space.

RealVision: You make the comparison to something like opioids in which there is a tremendous addiction problem, and you highlighted the idea that cannabinoids can represent a gateway away from this. When you look at this wellness fund, is this going to be a general theme though for you? Basically, taking things that would have been perhaps socially unacceptable, that may be neglected by traditional investors, and focusing on those. I mean this clearly is the immediate opportunity...

Harrison: This is a frontier market now. But I'll use my partner's story as an analogy.

When he was diagnosed with onset epileptic disorder, they gave him a drug to ingest, and then they gave him another drug to counterbalance the adverse effects of that first drug... straight down the line, he was stacked on six drugs and unable to function.

So one of the big benefits of cannabis in our view of the cannabinoid-based therapies is that the adverse effect profile is so de minimis. So you can get unstacked from those six medications through a single regimen if it's done the correct way, and if it's through the right lens, and through the right protocols. And that's why the FDA in the laboratory and the clinical trials is why we're focusing our attention on that specific subset.

RealVision: What are the benchmarks or markers that we can look at to see if this is projecting along the line? From an investor standpoint, what are some of the broad visible dynamics that we should be looking for? Is it further legalizations? Is it a federal process? Is the FDA reclassifying CBD? What are the markers that are more general us investors can use?

Harrison: It is a combination of things you're hearing and things you're not hearing. So for instance, Jeff Sessions ran this year-long trial to determine the negative impact of opioids and cannabis. Talked about how bad opioids are if you never mentioned cannabis once. Now if there was negative there, he would have jumped on that. So the fact that he didn't mention it, we view is as powerful as if he did come out and say something. Obviously, there's an agenda there.

But we're looking, again, at medical efficacy. So you have the GW NDA for Epidiolex, which is the epilepsy drug. You have the Glioma trials secondary endpoints for phase two... the longer that they take, the more positive it is. And then you have the election processes, and you have these other trials in the pipe. And the more efficacy that's demonstrated, I think it's creating a groundswell that's going to eventually overtake the current approach to cannabis.

RealVision: All right. Fantastic talk. Thank you again, and really enjoyed it.

Harrison: Thank you.

This interview was edited from its original form to help with comprehension.

To watch the original interview, visit www.RealVision.com.

PART II

Making a Fortune in Legal Marijuana
How the Millionaires Did It... and How You Can, Too

– Chapter 1 –

Marijuana Millionaires: How Regular Investors Are Cashing in on the Next Big Boom

The experts know there is a fortune to be made in the legal marijuana industry.

As Aaron Salz – the first Canadian equity analyst in cannabis – told RealVision.com...

> When you think about the global cannabis industry being a $100 billion, $200 billion-plus opportunity, there's certainly the ability for entrepreneurs and new business owners to really mint millionaires in this industry but also probably billionaires in this industry.
>
> We're just in the first inning of seeing who those business owners are going to be. But you can be certain that we'll have millionaires, and we will definitely have billionaires.

Every paradigm shift leads ordinary folks to the kind of wealth they never thought they could achieve. The coming marijuana boom will be no different.

In this chapter, we'll show you real world examples of people who have already transformed their lives and earned millions in this exciting new industry.

Before hitting their fortunes, the millionaires on this list were regular people spanning all walks of life. They were bankers, lawyers, venture capitalists, and event planners. They range from middle-aged mothers to young entrepreneurs.

The one thing they have in common is that they saw an opportunity in an emerging industry... They saw the shift in global attitude toward marijuana. They saw the long-term potential that shift could have. *And they didn't hesitate.*

These folks have all started businesses now worth millions of dollars. They've created jobs, paid taxes, and navigated the intense legal regulations some of our experts discussed earlier.

And today, they're reaping the rewards.

Of course, these are extreme examples. Just like any industry, not everyone who gets involved will strike it rich. But the following stories are further proof that legal marijuana is already booming. And the opportunities for even average investors could be extraordinary.

Now let's meet our millionaires...

Name: Hank Borunda

Location: Pueblo, Colorado
Business: Greener Side dispensary – grows and sells marijuana
Highlight: Made $1.5 million in first three months

> **❝** *I've known since I was a kid that I was going to make money.* **❞**

Hank Borunda's father always told him not to work for anyone else. Hank knew from an early age that if he really wanted to strike it rich, he'd have to work for himself. His road toward entrepreneurship started early, when he sold pine nuts to neighbors and passersby as a young boy in his hometown of Pueblo, Colorado.

But when Colorado's state legislature voted to legalize recreational marijuana in 2012, Hank pounced on the opportunity to enter the big leagues.

Hank was only 23 years old when he started the "Greener Side" dispensary in Pueblo. His mother initially opposed Hank's venture into the marijuana market. But she quickly realized the legitimacy of the business... "I listen to him talking to engineers and commissioners, and I'm in awe," she said.

It took the state two years to get the necessary rules and regulations in place to allow recreational sales. That gave Hank all the time he needed to make sure he was ready when the flood of new demand hit the market...

Colorado opened its doors for recreational marijuana sales on January 1, 2014. That day, Hank's dispensary was overrun with hundreds of customers. The line filled up his entire shop and spilled out the front door. He cleared $47,000 in 24 hours. Within three months, he grossed $1.5 million.

Few industries in the world today are growing fast enough for a 25-year-old kid to go from essentially nothing to owning a multimillion-dollar business within three months.

Hank used the proceeds to grow the business. Within six months, he had 20 employees and two retail outlets. He hired friends and neighbors, providing jobs and incomes for his community. He constructed a 3,600-square foot greenhouse to produce the marijuana, controlling his product and distribution from seed to sale.

Hank's father's advice was right all along... Working for himself allowed Hank to carve out his own slice of the American dream. And he did it all before turning 30.

Name: Michael "Big Mike" Straumietis

Location: British Columbia, Canada
Business: Advanced Nutrients
Highlight: On his way to becoming a marijuana billionaire

> **❝** *I actually have fun every day going to work.* **❞**

Michael Straumietis is a man by many names.

On Instagram, he's known as the "Marijuana Don." His photo stream is filled with scantily clad women lounging around any one of his several mansions, often indulging in marijuana consumption.

Others simply call him "Big Mike," due to his towering 6-foot 7-inch frame.

But it was a fake name – or seven – that laid the foundation for his marijuana nutrient empire...

Mike lost his father at age 17, and his mother was an alcoholic. But he credits a turbulent childhood for the inspiration to become an entrepreneur and create his own company. All it took were some friends explaining just how lucrative growing marijuana could be, and Mike was sold.

Only back then, in the early 1980s, it wasn't legal in most places. That didn't stop him.

At age 23, Mike devised hidden grow-ops in the crop fields of Illinois, as well as various indoor operations. His illegal business did not go unnoticed. Over the past three decades, Mike has been the object of a federal manhunt by the DEA and U.S. Marshals. He has lived on the run in different countries under seven different identities. He even served three months in jail.

But he is also a pillar of the now-legal marijuana industry.

If you've grown or consumed marijuana in the last decade, you've benefitted from the work of Mike Straumietis.

In his various operations, Mike has seen the cultivation of more than one million marijuana plants, introduced 50 revolutionary products and concepts to the field, and now focuses on plant nutrients and soil-less fertilizers through his company Advanced Nutrients.

Unlike most of his competitors, Mike specifically targets marijuana cultivation with his nutrient line. That decision, plus years of growing insights, has catapulted Advanced Nutrients into one of the leading nutrient suppliers in North America. Each year, Advanced Nutrients generates more than C$60 million in sales.

Mike has gone from outlaw to celebrity within the cannabis community. With his leading nutrient supply company, Mike makes millions each year from the industry without ever touching the plant. As the legalization boom spreads around the globe, there's no end in sight for this leading nutrient brand. He could become one of the first marijuana billionaires.

Name: Chloe Villano
Location: Denver, Colorado
Business: Clover Leaf University
Highlight: Made $2 million within four years

> *Growers were coming from all over the nation to meet me.*

The greatest tragedies can spark the greatest opportunities. For Chloe Villano, losing her younger brother to cancer marked a transformational moment in her life.

While searching for treatment options, she discovered the healing properties of cannabis for cancer patients. Even though her brother ultimately lost his fight, Chloe realized she could still help others with the medicinal aspect of the plant.

Chloe's background as a paralegal provided a unique opportunity to participate in Colorado's legal marijuana boom without ever touching the plant.

On June 28, 2010 – the anniversary of her late brother's birthday – Chloe set up Clover Leaf Consulting. The firm is designed to help marijuana entrepreneurs stay on the right side of the law while navigating a complex legal and regulatory land mine.

A self-described workaholic, Chloe has gone on to launch several successful offshoots from her first business. She launched Clover Leaf University in 2012, America's first state-licensed cannabis university.

At the young age of 33, Chloe earned $2 million within four years of starting her first cannabis business. Despite her meteoric rise in the cannabis industry, she stays humble. She drives a 10-year-old Mercedes. She donates to children's charities. And she pioneered a trade association to help marijuana entrepreneurs around the country get started in the industry.

Name: Nancy Whiteman
Location: Boulder, Colorado
Business: Wana Brands
Highlight: Created a $10 million company with a $30,000 investment

> *Whatever your stereotype might be of somebody in the marijuana business, I'm probably not it.*

Marijuana success stories come in all forms. Nancy Whiteman is a middle-aged mother and former marketing executive, and rarely indulges in enjoying the marijuana-infused candies she is famous for.

A self-described marijuana "lightweight," she's too busy expanding her edibles empire to enjoy the fruits of her labor.

She's working around the clock to become the first national edibles brand in America. By all accounts, she has the products and business smarts to make it happen.

Nancy and her now ex-husband John didn't have much experience with marijuana or running their own business before starting Wana Brands in 2010. But today, their business generates nearly $10 million in annual sales. Wana sells about a dozen different products, including neon-colored, marijuana-infused sour gummies. These gummies reportedly taste like "gourmet South Patch Kids" and have become Colorado's bestselling edibles product.

Nancy and John stumbled into the marijuana business when a friend acquired a commercial kitchen for making marijuana-infused sodas. They launched Wana not long after. And in 2015, they got their big break...

Denver-based EdiPure owned the top spot in the Colorado edibles market with its extremely popular sour gummy bears. But the company had trouble staying compliant with Colorado state regulations. In 2015, EdiPure recalled 63,000 products from store shelves.

Seeing the opportunity to fill a void in market supply, Wana invested $30,000 developing its own sour gummy recipe. Nancy hired a man she calls the country's top "gummy expert," and even ran the product through professional laboratory testing. Within six months of EdiPure's recall, Wana's sour gummy took the top spot as the No. 1 edibles seller in Colorado.

In 2018, the company's products dominate shelf space in the Colorado market. You'll find $20 canisters of Wana sour gummies in more than 450 of 500 dispensaries throughout the state. Nancy has plans to go nationwide. Keep an eye out for Wana Brands as the legalization movement grows. It could become the next big candy maker in America.

Name: Joel Schneider

Location: Denver, Colorado
Business: Bud and Breakfast
Highlight: Invested $100,000 into a business that now generates $110,000 a month.

> *I practiced law for 30 years and hated it... [Now] every day there's a different 420 happy hour.*

Joel Schneider was tired of the rat race. For most of his adult life, he'd worked as an attorney on Wall Street. He was successful, but he wanted to try something new.

So when Colorado legalized recreational marijuana in 2012, Joel traded in the concrete jungle of New York City for a 7,000-square foot, 19th-century home in Denver – his first "Bud and Breakfast" location.

Today, Joel owns three Bud and Breakfasts that cater specifically to marijuana users. Unlike traditional hotels, which forbid smoking of any kind, Joel's guests are encouraged to indulge in smoking marijuana. His suites come stocked with a full bar and serve meals catered by chefs.

"There is no place like this," Joel says. And his booking rate proves it. Most weekends are sold out, and guests book reservations for April 20 – the unofficial celebration day for cannabis users all over the world– a full year in advance.

With an initial investment of just $100,000, Joel's Bud and Breakfasts now generate $110,000 in sales *each month*. That's more than $1.3 million per year.

Few other asset classes offer that kind of return. For Joel, marijuana offered the ultimate freedom to break out of the nine-to-five routine and create his own reality.

Name: Cheryl Shuman

Location: Beverly Hills, California
Business: Beverly Hills Cannabis Club
Highlight: From $5 per hour to the owner of a billion-dollar business

> " *Growing up, I knew I was going to start a family business... I just didn't know it would be in pot.* "

Cheryl Shuman never envisioned herself as the "Cannabis Queen of Beverly Hills." At age 23, she was a single mother making $5 per hour at an eyeglass store in Encino, California.

But she always had the entrepreneurial streak...

She once turned a chance encounter with actress Shirley MacLaine into a multimillion-dollar business selling eyeglass frames and lenses to movie and television film sets for big hits like *Cheers* and *Terminator 2*.

In 2006, Cheryl was diagnosed with ovarian cancer. Hooked on a stream of powerful pharmaceuticals, including an IV morphine drip, she described herself as "bedridden, catheterized, and totally dependent."

Cheryl searched for alternatives to break the addiction and discovered medical marijuana. It helped her manage the pain without any of the debilitating side effects. Within 90 days of switching to cannabis, she was out of the hospital and working full time again.

Inspired by her own experience, Cheryl founded the Beverly Hills Cannabis Club in 1996 – a high-end dispensary famous for its celebrity clientele and exclusive pot parties. Today, she is one of the most high-profile cannabis entrepreneurs and activists in California.

Cheryl's marijuana dispensary is just the start. Through her holding company, Cheryl Shuman Inc., she has branched into other products, including the "Hautevape" vaporizer collection. These high-end cannabis-smoking devices come decked out with gold-plating and pave-set diamonds, with price tags in the five- to six-figure range. Her business is worth more than a billion dollars.

With marijuana going fully legal in California at the start of 2018, expect to see and hear more from the Cannabis Queen of Beverly Hills.

Name: Brooke Gehring
Location: Denver, Colorado
Business: FGS Inc.
Highlight: 100 Employees and more than $10 million in revenue

> **"** We knew it was worth jumping. What was the worst thing that could happen at 28 or 29 years old if we failed? **"**

In 2009, Brooke Gehring was stuck behind a desk, shuffling through the broken dreams of bankrupt American homeowners. She was a corporate banker during the depths of the financial crisis. But she found a silver lining...

As the whole world rushed for the real estate exits, one group was buying – marijuana investors. Her phone rang off the hook for marijuana entrepreneurs looking to buy distressed properties for their businesses.

Brooke wanted a piece of this booming market. She ditched her corporate banking job and launched a consulting service called LiveGreen Consulting. The company aimed to help entrepreneurs navigate the complex legal landscape of Colorado's burgeoning marijuana market.

Before long, one of Brooke's clients asked if she wanted to purchase his marijuana cultivation and dispensary licenses. The lengthy approval times required for these licenses makes them valuable commodities. So she made the leap from advising others to actually running a marijuana growing and selling business.

The early days contained a mix of excitement and turbulence. She lost an entire warehouse of plants when her lease owner defaulted in 2011. Strict federal banking laws caused her to lose over 30 bank accounts. But in the end, Brooke's perseverance paid off in a big way...

In 2018, she oversees two thriving cultivation sites, comprising 35,000 square feet of growing space in Denver, Colorado. These operations supply her three dispensary locations, which distribute marijuana products for both medicinal and recreational use. She has about 100 employees and generates more than $10 million in annual revenues.

Name: Darren Roberts

Location: Boca Raton, Florida
Business: High There
Highlight: Gained 600,000 users within first three years

> **"** *We've been called Facebook for cannabis.* **"**

In today's world, every new trend demands a social media presence. The cannabis boom is no different.

Darren Roberts and Kenneth Frisman capitalized on this opportunity by creating an online meeting place for cannabis users – the "High There" social media app.

The app brings together fellow cannabis connoisseurs in a tinder-like interface, where they flip through the profiles of different members and swipe left or right if they'd like to connect with one another.

Within three years, the app grew from zero to more than 600,000 active users. One couple who met on the app ended up getting married. Though the founders describe High There not as a dating app, but as a "meet people make friends" app.

Another user with terminal cancer turned to the app for advice on consuming marijuana in her final months of life. "As sad as it was," Darren said, "it was really cool to know that you had the power to connect people."

Darren and Kenneth were both in their mid-40s when they launched High There. Neither had any experience in cannabis or social media apps – Darren spent the previous 15 years in real estate. Kenneth managed public equities via his family office.

So how did they break into an industry they knew nothing about and actually succeed? Darren credits old-fashioned hard work for their success. He explains that while the industry is new, the same time-tested principles of business still apply...

> At the end of the day, it comes down to the basic principles of business. You have to see how you can go ahead and look to generate revenue and what your ROI will be, cash flow needs, and so forth.

The app is free now. And based on user reviews, it has some technical issues that need to be worked out. But based on similar industry valuation metrics of $120 per user, the 600,000-member base gives Darren's and Kenneth's company an estimated $72 million valuation.

As the marijuana boom goes global and brings this once taboo industry into the mainstream, the future for cannabis-based social networking looks bright.

Name: Isaac Dietrich

Location: Denver, Colorado
Business: MassRoots
Highlight: Transformed $17,000 into a $55 million business

> **"** *We really want to be the first billion-dollar cannabis technology company. That's our main objective.* **"**

In April 2013, Isaac Dietrich and Tyler Knight hatched the idea of a social networking company for cannabis users. Their reasoning: "They didn't want their grandmothers to see pictures of them taking bong rips on Facebook."

Isaac initially had trouble convincing Silicon Valley to get behind the idea. He was turned down by 150 people. But that didn't stop him. Isaac describes what he did next...

> I ended up taking out $17,000 in credit card debt to get the company off the ground. My parents nearly killed me, but that's part of the fun.

It took only a few months to go from idea to product. The MassRoots social networking app launched in the Apple App store in July 2013. By March 2014, the application's user base was growing by 20,000 to 30,000 people per month.

By the end of that year, MassRoots ranked as one of the fastest-growing companies in the App store. The company went public in April 2015 under the ticker MSRT.

The company has had its share of troubles in the last couple years – racking up millions of dollars in losses, defaulting on its debts, laying off 40% of its workforce, and shuffling its board and management team. But the app continues to grow. And as of early 2018, MassRoots has more than one million users.

Name: Ralph Morgan

Location: Denver, Colorado
Business: O.penVAPE
Highlight: From no experience to $100 million company

> **"** *We're creating a product that is three to five times the value of gold.* **"**

Ralph Morgan's success story should inspire every budding entrepreneur. He started with almost no marijuana experience and became the CEO of one of the largest privately-owned cannabis companies in the U.S. He did it all in less than 10 years.

Before breaking into the cannabis space, Ralph previously worked as a health care professional. His background and research convinced him of the tremendous potential in medical marijuana.

What he lacked in experience, Ralph made up for with entrepreneurial instinct. After opening his first dispensary – Evergreen Apothecary – in 2009, he quickly recognized an unmet need in the market: consistent, healthy, and organic vehicles for consuming marijuana.

He foresaw vaping as the next major innovation to meet that need. Vaping involves heating THC-concentrated oils and inhaling a relatively pure vapor. Many prefer this over burning marijuana flowers, which can irritate the throat and lungs.

To capitalize on this trend, Ralph launched a new company, Organa Labs. Here, he explored the best way to extract cannabis oil from the plant flowers (carbon dioxide). He also created a new brand of vaporizer pens used to consume cannabis oil – the O.penVAPE brand.

O.penVAPE CFO Steve Berg describes Ralph's foresight...

> [Ralph Morgan] was the early pioneer in terms of seeing the importance and opportunity in cannabis oil, specifically [carbon dioxide]-extracted cannabis oil.

That foresight paid off in a big way. It helped Ralph's O.penVAPE products become the No. 1 seller at dispensaries around the country. The company did about $100 million in sales in 2016, and marijuana consumers around the country buy its products every 10 seconds. Comedian Sarah Silverman was spotted puffing one of the company's O.penVAPE pens on the red carpet at the Oscars.

Ralph's stunning success story shows that anyone can become the next marijuana millionaire – no experience needed.

Name: James Howler
Location: Denver, Colorado
Business: Cheeba Chews
Highlight: From home kitchen to America's leading brand of cannabis candies

> **"** The business is completely self-sufficient from a financial perspective. No debt, low overhead, efficient processes... and we've taken no outside investment. **"**

All aspiring marijuana entrepreneurs should take a note from James Howler's book. He started experimenting with marijuana edibles in his personal kitchen back in 2009. That hobby turned into a booming business.

And less than 10 years later, his Cheeba Chews marijuana-infused candy business sells more than 50,000 products each week in 800 dispensaries.

But here's the best part – he did it all without taking on a dime of debt or outside investor capital. The company remains 100% privately held and self-funded. Here's how it all got started...

James was active in the Colorado medical marijuana scene early on. He already had his growing license by 2009. And he noticed one major flaw in the edibles side of the industry – inconsistency in the potency of products. Not knowing exactly how much THC was in any given product made consumers wary and hampered sales.

So he figured out a way to achieve consistent dosing in bite-sized chews, using a premium cannabis oil. To ensure consistency, the oil is tested at three different stages – in the flower buds after harvest, in the extracted oil from those buds, and finally in the edible itself.

This focus on consistency has made Cheeba Chews America's leading cannabis-infused candy brand. It has won three Cannabis Cup competitions – the marijuana industry's most widely esteemed competition.

By keeping the business completely privately owned and self-funded, Howler weathered the inevitable ups and downs of growing a business from scratch, including the time the company lost about two weeks' worth of inventory when one of the kitchens temporarily shut down.

James Howler shows that focusing on the details separates the good brands from the great. And disciplined financial management can keep your company afloat even when inevitable turbulence strikes.

Name: Nick Kovacevich
Location: Santa Cruz, California
Business: Kush Bottles, Inc.
Highlight: Growing sales at 250% per year

> **"** We see our company as... going to benefit, regardless of the direction or who the real winners are in the industry. **"**

Nick Kovacevich's father – a former assistant district attorney and judge in Santa Cruz, California – used to spend his days locking people up for possessing marijuana.

These days, his son makes a fortune... legally... in the marijuana industry. And he does it without ever touching the plant.

In 2010, Nick co-founded Kush Bottles as a picks-and-shovels opportunity. Just like Levi Strauss during the gold rush, Nick understood that by providing the necessary packaging and accessories to folks who sell and consume marijuana, he could win regardless of who produces the most product.

Kush wholesales marijuana packaging materials to dispensaries around the country. It's perhaps best known for its single-roll tubes, used to hold joints (marijuana cigarettes). The company sells more than two million of these tubes per month, and it owns a patent on their child-resistant configuration. This patent could help the company secure a competitive edge and increasingly dominate the market.

Kush Bottles also sells a wide variety of other packaging materials, including vaporizer cartridges, smoking accessories like water pipes and vaporizers, marijuana grinders, and rolling papers. And it works with retailers to create customized, branded product packaging. All told, the company has more than 2,000 unique product offerings.

This diverse product lineup translates into rapid sales growth. In the last two quarters, Nick's company grew sales more than 250% year-over-year. Wall Street analysts expect it to generate nearly $50 million in annual revenues for 2018.

– Chapter 2 –

The Potential Winners of the Global Marijuana Boom

The opportunity in legal marijuana isn't just for entrepreneurs, like the newly minted millionaires we profiled in the last chapter. This industry is just taking off on a global scale.

If you want the best chance to win big in the global marijuana boom, the choice is easy: Start in Canada.

The country has cemented a huge lead among all potential competitors and offers the most straightforward operating environment for cannabis companies and investors.

Opportunities will emerge throughout the marijuana value chain, from seed to sale. I'm talking about the growers, distributors, and equipment providers, to name a few.

Marijuana legalization will move billions of dollars from underground black markets into the real economy.

However, with opportunity comes plenty of risk. For every major success story, we'll likely see a dozen or more failures and frauds. Given this backdrop, we've broken down our review of cannabis companies into the following categories:

- Majors
- Minors
- Picks and Shovels

: None of the companies we discuss in this
active recommendations.

If you choose to invest in the legal marijuana trend on your own, these are a few names you may want to consider.

But before making any decisions, do your own research. Draw your own conclusions. And invest only as you see fit.

Majors

As with commodities, experts believe that cannabis companies with the largest economies of scale – growing production and decreasing costs – will dominate the market.

In 2018, Canada is the only country that has multibillion-dollar publicly traded cannabis producers.

Medical marijuana is legal at the federal level in Canada. And the government is expected to legalize recreational marijuana around July 2018. A 2017 Ernst and Young report showed that 87% of Canadian producers believe industry consolidation is coming, "leaving a few large players post-legalization."

So far, three companies have emerged as those largest players – the so-called "Big Three" of Canadian pot. They are......

1. Canopy Growth Corporation (TSX: WEED)

2. Aphria (TSX: APH)

3. Aurora Cannabis (TSX: ACB)

Canopy Growth Corporation (TSX: WEED)

Canopy was the first cannabis grower to list on the Toronto Stock Exchange and the first North American company to expand internationally. Today, it's the clear leader in the Canadian medical-marijuana market... the uncontested economy of scale, producing the highest volumes of cannabis at the lowest cost.

Formerly known as "Tweed Marijuana Inc.," Canopy began operations in 2013 and made its first sale of medical marijuana in May 2014. It served 16,000 registered patients by mid-2016. By mid-2017, its client base had grown to 59,000 patients. That's nearly 270% growth in just one year.

In 2017, Canopy had the largest legal marijuana-production operation on the planet. Its flagship production site – 500,000 square feet on 40 acres of land in Smith Falls, Ontario – is the former headquarters of candymaker Hershey's Chocolate.

The company is targeting aggressive growth in production capacity to maintain its current leadership position...

In October 2017, Canopy entered into a joint venture with an experienced greenhouse operator to develop 1.3 million square feet of greenhouse growing capacity in British Columbia, with an option to develop another 1.7 million square feet of growth space.

In total, the company plans to boost capacity to more than five million square feet in the coming years. That's a more than tenfold increase in production footprint.

Canopy is also leading the charge into specialty branded medical products via partnerships with rapper Snoop Dogg and Corona beer maker Constellation Brands (STZ)...

In late 2016, Canopy partnered with Snoop Dogg to sell three specialty pot varieties under the brand "Leafs By Snoop." Leafs By Snoop cannabis sells for between C$9 and C$12 per gram versus an average industry price of around C$8.

The company also partnered with leading cannabis breeder DNA Genetics and distributor Organa Brands to develop proprietary cannabis seed strains.

Plus, Canopy introduced new derivative products, like cannabis oils and oil-based softgel caps.

These efforts have paid off in the form of higher sales prices per gram.

Canopy reported an average sales price across all brands of C$7.99 per gram for the third quarter of 2017 – up 14% from a previous average price of C$7.01 per gram in the third quarter of 2016.

Canopy is also working with Constellation Brands to develop, market, and sell a cannabis-infused beer that will go to market if recreational use in Canada is legalized (as expected). On October 30, 2017, Constellation purchased a 9.9% equity stake in Canopy.

These partnerships with industry leading brands could prove a major competitive advantage as traditional cannabis herb becomes an increasingly commoditized product. Specialty branded product offerings will provide Canopy the opportunity for higher margins and greater customer loyalty.

Its expanding product offerings will also help it become a major player outside North America, in the growing international cannabis market. As of early 2018, Canopy is a leading exporter into the German and Brazilian markets. It partnered with Australian grower AusCann to supply Australia with medical marijuana. And it's pursuing production facilities in Chile, Jamaica, and Denmark.

Canopy's low production cost, international expansion, and growing portfolio of branded products make it a leading candidate for the top spot in the global cannabis industry.

Aphria (TSX: APH)

As of early 2018, Aphria is the No. 2 publicly traded cannabis producer in Canada. But it's angling to overtake Canopy for the No. 1 spot with "the largest fully funded production capabilities in the industry in 2019."

Based in Leamington, Ontario, Aphria's production strategy is guided by former greenhouse operators Cole Cacciavillani and John Cervini. These industry experts helped Aphria grow from zero production in 2014 to one of Canada's largest producers within just three years. The company produced 1,237 kilograms (kg) in the quarter ending November 30, 2017. But this is only the start...

Aphria plans to grow its production to 220,000 kg per year by mid-2019. That's a more than twentyfold from annual 2018 estimates. Global equity-research firm Canaccord Genuity estimates Canadians will consume around 400,000 kg of cannabis per year by 2021. *This means Aphria's expected production volume could satisfy more than half of Canadian demand.*

This growth will come from a combination of organic expansion, joint ventures, and acquisitions. On the organic-growth side, Aphria has a four-stage plan in place to boost capacity 1,000% to 100,000 kg per year by 2019. The company will achieve this using one million square feet of production space.

In January 2018, Aphria formed a joint venture – "GrowCo" – with greenhouse grower Double Diamond Farms. Double Diamond will supply the land, greenhouses, and infrastructure to install capacity for up to 120,000 kg of annualized cannabis production. That gets Aphria's total expected annual production up to 220,000 kg.

The project will require an estimated C$80-C$100 million in financing. Much of that will come from smart acquisitions...

Aphria shares jumped around 350% in the second half of 2017. It capitalized on this soaring stock value to raise capital, with a C$115 million deal in December. Despite having the smallest market cap among the Big Three, this capital infusion arms the company with a sizeable war chest going forward.

In addition to its growing success in Canada, Aphria is eyeing expansion in the massive international cannabis market. On November 23, 2017, Aphria received its dealer's license from Health Canada – a permit that allows the company to export cannabis into international markets. Aphria is currently targeting the medical-marijuana markets in Australia, Italy, Germany, and Argentina.

These international exports will help ensure Aphria finds buyers for its massive production growth in the years ahead.

Aurora Cannabis (TSX: ACB)

Among the Big Three Canadian cannabis leaders, Aurora is the youngest. Based in Vancouver, British Columbia, Aurora made its first sale in January 2016. But don't let this late start fool you... In less than two years, Aurora grew to more than 20,000 active registered patients and C$2.5 million in monthly cannabis sales. According to management, this is the fastest patient growth in the industry.

Aurora's aggressive expansion plans threaten to shake up the race for Canada's top producer.

Aurora produces cannabis from two facilities in Alberta and Quebec. The 55,200-square-foot Alberta facility uses artificial lighting and hydroponics (growing plants in water-based mediums, without soil) to produce up to 5,400 kg of cannabis per year. Starting with its first harvest in 2018, the new 40,000-square-foot Quebec facility is expected to produce up to 3,900 kg of cannabis per year.

Aurora's next major expansion plan revolves around a massive new facility more than 10 times bigger than the Alberta and Quebec facilities combined – an 800,000-square foot grow site called Aurora Sky.

Built on land near the Edmonton International Airport in Alberta, Aurora Sky will produce up to 100,000 kg of cannabis per year at full capacity. Aurora management believes this will be the "most technologically advanced cannabis facility in the world."

Like its competitors, Aurora also wants a piece of the burgeoning international market for medical marijuana. In May 2017, it acquired Pedanios GmbH – the largest distributor of medical marijuana in Europe.

In August 2017, it obtained European Union GMP (good manufacturing practices) certification to export medical marijuana into the German market. Aurora made its first shipments of cannabis to Germany in September 2017.

On January 1, 2018, Aurora received a license to grow and sell cannabis in Denmark. It also signed a deal with Danish greenhouse tomato grower Alfred Pedersen & Søn to form joint venture Aurora Nordic, based in Odense, Denmark. Aurora will own 51% of the venture.

Aurora Nordic plans to construct a one million-square-foot production facility capable of producing 120,000 kg per year. It will use this capacity to sell cannabis in Denmark, Sweden, Norway, Finland, and Iceland.

Among the Big Three, Aurora offers the greatest growth but arguably the greatest risk. The company's price-to-sales (P/S) ratio – a good valuation metric for high-growth businesses – comes in at 120 as of February 2018. That's nearly double Canopy's P/S ratio of 63 and Aphria's 72.

This means that as of early 2018, the company trades at a very high valuation relative to its peers, suggesting a greater risk for investors if anything should go wrong.

The company was also flagged in January 2018 by noted short seller Citron Research for questionable accounting methods.

Of course, this doesn't mean Aurora can't be a good investment.

Again, we aren't recommending any of these companies. We urge you to do your own due diligence on all the businesses discussed in this chapter before deciding what investment – if any – is right for you.

Minors

Despite the trend toward industry consolidation, there will be opportunities for smaller players to dominate niche markets or to serve as potential acquisition targets for the Big Three.

The main companies you may want to pay attention to in this category are...

1. MedReleaf Corporation (TSX: LEAF)
2. Cronos Group (TSX V: MJN)
3. Village Farms (TSX: VFF)

MedReleaf Corporation (TSX: LEAF)

MedReleaf is a mid-sized producer in Ontario with a focus on innovating new cannabis-treatment solutions.

Armed with a staff of PhDs, the company experiments with more than 200 strains of cannabis to find tailor-made medical treatments. The company has developed the first "cannabis compatibility test" targeting specific metabolic pathways for individual patient solutions.

Incorporated in 2013, MedReleaf was one of the original medical-marijuana producers in Canada. Five years later, it's still a leader in operational excellency. MedReleaf is the world's first cannabis company with four internationally recognized certifications, including ICH-GMP and ISO 9001 (Quality Management System) licenses.

This operational excellence shows up in the company's industry leading production efficiency. **While it is not a leader in total volume of production, it is a leader in maximizing its production space.** In the third quarter of 2017, MedReleaf reported "best-in-class cultivation yields" of 300 grams per square foot per year. **Those are the highest yields per square foot in the industry.**

As of early 2018, MedReleaf operates the Markham Facility only, which produces up to 6,000 kg of cannabis and 1,760 kg of cannabis oil per year. It plans to grow capacity with its new Bradford Facility – a 210,000 square foot space with production potential of 28,000 kg of cannabis per year. The first phase build-out was completed in April 2017 and has produced several successful harvests.

MedReleaf is also a leader in customized treatment solutions. It owns 15,000 seeds and more than 200 unique strains of cannabis. The company's PhD-led genetics department oversees the breeding process aimed at delivering specialized medical relief targeting the individual needs of its patients.

And MedReleaf innovates novel solutions for delivering cannabis to patients via oils, gels, lotions, and more. The reason is simple... Industry reports suggest that cannabis extracts like oils represent about half of total cannabis sales. MedReleaf entered the cannabis-extract market in late 2016. By the third quarter of 2017, the company generated 18% of total revenues from cannabis-extract products.

MedReleaf was the first licensed Canadian producer to create a cannabis-infused topical cream in October 2017. The cream was developed in response to patient feedback and demand for topical applications of CBD. It's designed for optimal absorption with MedReleaf's CBD-focused cannabis strains.

The company also just released a new product for customized patient delivery. ReleafDX is "the first pharmacogenetics-based cannabis compatibility test" from a licensed producer in Canada. The test uses a cheek swab to analyze biomarkers that will help physicians design specific dosing and strains targeting optimal metabolic pathways for individual patients.

MedReleaf is also pursuing international growth, including partnerships with Israeli and Australian cannabis firms.

Cronos Group (TSX V: MJN)

Cronos Group started out in 2012 as a holding company designed to invest in Canada's nascent medical marijuana industry. The company began acquiring top-tier assets in 2013, creating a foothold in the industry before the boom sent valuations sky-high in 2016-2017. Smart investments have propelled the company from simple investor to an owner and operator of marijuana production facilities across the globe.

These days, Cronos invests in and operates a diverse line of medical marijuana companies. The investment portfolio contains privately held Canadian producers, including a 21.5% equity stake in Whistler Medical Marijuana and up to a 30% stake in Evergreen Medicinal Supply.

Cronos also owns shares in two publicly traded companies – AbCann Global (ABCN.V) and Big Three industry leader Canopy Growth (WEED.TO). As these companies do well, so will Cronos.

The company's flagship production facility in Ontario – Peace Naturals – provided the platform for organic growth. The 95-acre facility contains 40,000 square feet of production space across three production buildings. It can generate 5,000 kg of cannabis per year.

To expand capacity, Cronos is constructing a fourth massive 286,000-square-foot building on the site – named "Building 4" – that can produce 33,500 kg of cannabis per year. Production from Building 4 will come online in the second half of 2018.

Cronos' Peace Naturals operation is one of the first Canadian production facilities to have earned the coveted GMP certification in Germany. And it wasted no time capitalizing on this key license...

In October 2017, Cronos entered a five-year exclusive partnership with pharmaceutical manufacturer Pohl-Boskamp to distribute Peace Naturals cannabis into Germany. This will provide a major opportunity, given Pohl-Boskamp's sales presence in more than 12,000 German pharmacies.

Cronos is pursuing further international growth with a joint venture in Israel under the name "Cronos Israel." Cronos partnered with Gan Shmuel – an agricultural collective (or "kibbutz") in northern Israel, which is in the process of obtaining a cultivation license from Israel's Ministry of Health.

In Phase I of the Cronos Israel project, the company will build a 45,000-square-foot greenhouse that will generate 5,000 kg of cannabis per year. Phase II will boost capacity up to 24,000 kg per year.

Situated on more than 4,900 acres, Gan Shmuel provides the opportunity for further expansion beyond Phase I and II, with the potential to grow 100,000 kg of cannabis per year. Gan Shmuel will contribute its large footprint of existing agricultural infrastructure, on-site water access, and pool of skilled agricultural labor.

The best part of this agreement is the world-leading operating climate – abundant sunlight and just the right amount of humidity. Due to Israel's favorable growing climate for cannabis, Cronos Israel projects a meaningful reduction in production costs – spending as little as C$0.40 per gram in Israel compared with C$2.20 in Canada.

The combination of international export opportunities, a diverse investment portfolio, and a rapidly growing production footprint could make Cronos a prime acquisition target. Alternatively, Cronos may seek to merge with another mid-size producer to grow its scale and compete with its larger rivals.

Village Farms International (TSX: VFF)

British Columbia-based Village Farms International is one of the largest greenhouse vegetable growers in North America.

In 2017, the company began its expansion into cannabis production. As of early 2018, the market is ignoring Village Farms. It isn't included in any of the big marijuana-based exchange-traded funds (ETFs). *But once it starts actually producing the stuff, Village Farms could quickly become one of Canada's largest cannabis producers.* That's because it has an edge...

Village Farms currently operates 10.5 million square feet of greenhouses growing vegetables. It generates hundreds of millions in revenue each year. Compare this with current cannabis production leader Canopy, which boasted about its "over 2.4 million" square foot facilities under development in late 2017.

The size of Village Farms' greenhouse footprint and years of growing experience are unprecedented in the cannabis industry...

Nearly three decades of success operating as one of the lowest-cost growers of greenhouse tomatoes and peppers means Village Farms can compete with the majors in today's marijuana market. Village Farms believes it can produce cannabis at an all-in cost of less than C$1 per gram. This compares with an average cost of C$2.20 for the four largest producers (Canopy, Aphria, Aurora, and MedReleaf).

Because of its low production costs, Village Farms expects to lock in a higher profit margin by producing cannabis than by producing vegetables. Increasing margins from single-digit territory with just vegetables to more than 50% with cannabis could boost revenues 10-15 times.

Village Farms currently owns seven greenhouses. Three of these are clustered together in British Columbia and contain nearly 5 million square feet of production capacity (about half of the company's total).

In June 2017, Village Farms and licensed Canadian cannabis producer Emerald Health Therapeutics (EMH.V) formed a 50/50 joint venture called Pure Sunfarms. Pure Sunfarms will convert one of the three British Columbia greenhouses from vegetable to cannabis production. Emerald Health will contribute the capital needed for conversion. Village Farms will provide the facility, existing infrastructure, and a rich labor pool.

Village Farms' immediate goal is to produce 75,000 kg of cannabis per year by 2020 with the first greenhouse conversion. Longer term, it hopes to convert all three greenhouses and estimates production at a conservative 300,000 kg annually. This could make it one of the largest (if not *the* largest) legal cannabis producer in the world.

On the surface, Village Farms looks like a regular vegetable grower. It's arguably priced like one. But once the market realizes it has the potential to become one of the top marijuana producers in Canada, the share price should surge.

That and its hundreds of millions in existing revenues could garner it a multibillion-dollar valuation like its future peers Canopy, Aphria, and Aurora.

Picks and Shovels

Typically, investors look to play booming commodities markets by investing in commodity producers, like the major and minor players we've already discussed in this chapter. That can be a lucrative but tough game.

Producers take the big risks for the big payoffs... They can generate incredible growth, like Aphria's plans to boost production twentyfold in 18 months. If successful, this surge in production could generate massive revenue and earnings growth... ultimately rewarding investors with a rocketing share price. But this kind of growth isn't cheap and adds an element of risk to an investment in these companies...

Commodity producers have hefty capital requirements. If anything goes wrong, a lot of shareholder capital can get wiped out. Plus, they can face macro headwinds from competitors, which they have no control over. Rapid growth from competitors can put downward pressure on margins as supply floods the market.

Sometimes, the most – and safest – money is made with "picks and shovels." These companies provide vital equipment and services that the producers need to produce their product.

The classic picks and shovels success story is set in the 1850s...

Back then, a German immigrant moved from New York to San Francisco to participate in the California Gold Rush. Rather than the "all or nothing" route of looking for a big gold strike, though, this guy sold basic goods to the miners. He eventually started producing a new type of durable pants. They became a huge hit... And he got rich.

His name was Levi Strauss. Strauss didn't risk it all on trying to find the big strike... He just sold the stuff everyone else needed to try to find the next big strike themselves.

The idea of owning picks and shovels has become an investment cliché for good reason. It can be an incredibly profitable, diversified way to profit from rising commodity prices.

Our research has identified one specific picks and shovels name that stands out in the legal marijuana industry...

Scott's Miracle-Gro (NYSE: SMG)

Scott's Miracle-Gro is a household name in America.

The company sells some of America's leading fertilizers and pest control brands, including Miracle-Gro, Ortho, and Roundup. These products dominate the lawn and garden shelf space at major retailers like Home Depot, Lowe's, and Walmart.

Scott's core business is highly profitable. It consistently generates gross margins above 30%. But with U.S. home ownership near multidecade lows, the domestic lawn and garden market offers little growth potential. And those margins have remained stagnant since 2008.

The booming cannabis market has provided a new a shot in the arm for Scott's...

When you think of growing marijuana, you might think of rolling fields like this...

But today's sophisticated producers don't grow their product in the ground. They grow year-round in high-tech facilities, often using hydroponics bays like this...

As we said earlier, hydroponics is the practice of growing plants in soilless mediums, where the plants' roots are suspended in nutrient-rich water-based solutions. This provides more efficient nutrient uptake than traditional soil growing, resulting in faster growth and higher volumes.

Cannabis producers can sell their product for up to $4,000 per pound as of early 2018. So they spare no expense to squeeze every ounce of growth from their operations.

On average, U.S. households spent about C$400 each year on the traditional lawn-care products in Scott's core business. Shoppers in hydroponic stores, however, can spend that much in a single trip – often in cash.

Since 2014, Scott's has spent half-a-billion dollars acquiring leading brands of indoor hydroponic gardening equipment. These include hydroponic nutrient leader General Hydroponics, indoor growing lightmaker Agrolux, and the leading indoor air filter maker in Canada Can-Filters Group.

Scott's began breaking out the performance of the company's hydroponics unit – named "Hawthorne" – in 2016. From the company's fiscal 2016-2017 years, Hawthorne sales grew 137%. Even better, profits for the division grew 201%. Much of this growth came from acquisitions, but the underlying hydroponics businesses generated organic sales growth of 23.2%.

The best part is, Scott's doesn't depend on legalization gaining further traction in the U.S. to keep making money. Whether pot is sold on the black or white market, growers still need equipment. Scott's is going to supply them.

– Chapter 3 –

Get Your Degree in Smart Trading Basics: Three Ways to Manage Risk

To become a rich, successful investor, you must always focus on how you can *lose* money. You must always focus on risk – and how to avoid it.

Once you've taken care of the risk, you can move on to the fun stuff... *making* money.

The thing is, amateur investors are always 100% about the upside. They're always thinking about the big gains they'll make in the next big tech stock or their uncle's new restaurant business... or the legal marijuana boom.

They don't give a thought to how much they can lose if things don't work out as planned... if the best-case scenario doesn't play out.

And the best-case scenario usually *doesn't* play out. Since the novice investor never plans for this situation, he gets killed.

That's why you must always plan for the worst-case scenario. It will minimize your risk, maximize your upside, and make you a vastly better investor.

There are three main ways to do this...

1. Decide on your asset allocation.
2. Calculate your position size.
3. Define your exit strategy.

Learn these techniques – and practice them – like a pro.

As long as you have your "degree" in smart trading basics, you can invest with the confidence that you won't blow up your portfolio.

Asset Allocation

The single most important factor in your investing success has nothing to do with picking the right stocks.

It has nothing to do with paying attention to what the politicians say or knowing how to time the market.

The single most important factor in your investing success is 100 times more important than any of those things.

Ignorance and mismanagement of this factor ruin more portfolios than every other factor combined. Yet most investors never give this idea any thought...

This vitally important idea is called "Asset Allocation."

Think of asset allocation (how you divide your investment dollars into stocks, bonds, etc.) as you would your diet. A diet of all meat could literally kill you. A diet of all vegetables, while it may not kill you, would likely lead you to inferior overall performance. We all know this. And we know that some balance of these is optimal – hence the phrase "well-balanced diet."

The same holds for investing. We know that some mix of stocks and bonds over the long run is the "optimal" mix to give us healthy gains and (at the same time) ward off portfolio heart attacks. Let's go over the details...

<u>Asset Allocation Accounts for 90% of Portfolio Performance</u>

Asset allocation is your mix of stocks, bonds, and cash.

Now-famous studies (by Gary Brinson in 1986 and 1991) show that more than 90% of portfolio performance and variability can be explained by asset allocation. The other 10% is made up of factors including market timing and stock selection.

This makes sense. If your asset allocation is 100% stocks and the market falls 20%, chances are you'll be down around 20% – regardless of the stocks you choose. If your allocation is 100% bonds and bonds earn 6%, chances are your bond portfolio will make you 6%.

Clearly, the most important decision you make is how you divide up your pie of assets. If those studies about portfolio performance are even remotely correct (they do intuitively make sense), it means that picking stocks isn't the best use of your time – because a rising tide raises all ships. (And, yes, unfortunately, an outgoing tide has the opposite effect.)

How to Figure Your Own Personal Asset Allocation

Many people have no idea what sensible asset allocation is... So they end up taking huge risks by sticking big chunks of their portfolios into just one or two investments.

For example, consider employees of big companies who put a huge portion of their retirement money into company stock. Employees of big companies that went bankrupt – like Enron, WorldCom, Bear Stearns, and Lehman Brothers – were totally wiped out. They believed in the companies they worked for, so they put more than half of their retirement portfolios into company stock.

And it's all because they didn't know about proper asset allocation. Because of this ignorance, they lost everything.

Keeping your wealth stored in a good, diversified mix of assets is the key to avoiding catastrophic losses.

If you keep too much wealth – like 80% of it – in a handful of stocks and the stock market goes south, you'll suffer badly. If you're heavy in real estate (like many folks were in 2006), you'll be wiped out in a big real estate crash (like many folks were in 2008).

The same goes for any asset... gold, oil, bonds, real estate, or blue-chip stocks.

You can get an entire degree learning how to "optimize" these in your portfolio... trying to nail the perfect risk/reward ratio. Heck, folks have studied the stuff in-depth and done the math behind asset allocation and portfolio theories and won the Nobel Prize for their work.

But much of the "optimization" the geeks come up with is based on forecasts and past data that are wrong more often than they are right. You can get ahead of 90% of investors with a much simpler process.

First, **you start with allocating a little pile of money into an emergency fund**... You should keep it liquid as cash in a savings or checking account. It makes sense to put aside enough to last you between three and six months, depending on your personal need for safety.

Once you set aside some cash for emergencies... start with a simple allocation where you **decide between just stocks and fixed-income types of securities (bonds)**. If you have a longer-term view and a higher tolerance for risk, you could make your allocation 80% stocks and 20% bonds. If you are closer to retirement and don't like volatile returns, you could do the inverse, 80% bonds and 20% stocks. Most of us fall somewhere in between.

The point is to select assets – like stocks and bonds – that are not perfectly correlated, meaning their price movements aren't tied to each other. Combining them in your portfolio will smooth out your overall returns.

You can easily get more complex, dividing your categories (or "allocations") into domestic and international stocks, corporate, government, and municipal bonds, and so on. You can even add a small allocation to precious metals.

But before you get into all that, start simple.

You can choose, say, a 60% stock and 40% bond allocation, and stick to it. A 60% stock and 40% bond allocation is a great "middle of the fairway" asset-allocation plan. It ensures you harness the proven wealth-building power of stocks... while also using the conservative, income-producing power of bonds.

Each year, with just a few hours of work, you should rebalance your portfolio. Be sure to do it every 366 days so you don't incur a tax bill on short-term gains.

As you get closer to retirement, you can adjust it to match your risk tolerance. For example, you can use the "60/40" asset allocation while you are in your 40s, 50s, and 60s... and then start increasing your allocation to more bonds and fixed-income types of investments when you reach your 70s.

There's no way anyone can provide a "one size fits all" allocation. Everyone's financial situation is different. Asset allocation advice that will work for one person can be worthless for another.

You've got to find a mix of assets that is right for you... that suits your risk tolerance and your station in life. Whatever mix you choose, just make sure you're not overexposed to an unforeseen crash in one particular asset. This will ensure a long and profitable investment career.

Position Sizing

As you can see, investing isn't a one-size-fits-all process. Everyone does it for different reasons.

And your personal goals have a lot to do with how you decide to manage your investments.

You could be looking to earn enough money in the stock market to send your child to college in 10 years. Or you might want to position yourself to make huge gains from a major macro event that you believe is on the horizon.

Maybe you want to protect yourself against economic calamity. Maybe you just want to continue to generate steady income to maintain your current lifestyle.

Your decision-making process and the risk-management strategies you use will be completely different in each scenario. It's important to realize that your goals matter. The more you understand your own goals, the more likely you are to achieve them.

And consistently achieving the goals you set will make you a successful investor...

Let's face it, no stock purchase will ever be risk-free... Taking risks gives investors opportunities to succeed. But different stocks have different levels of risk... And those levels can vary dramatically. So how much risk is the right amount? Again, there is no one-size-fits-all answer. It all depends on *your* level of risk tolerance.

Risk tolerance is how much exposure to loss you're comfortable with. It's how much you can afford to lose in pursuit of a big payoff... and how long you can wait to get paid.

The amount of risk you take can determine how quickly you meet your goals... or whether you even meet them at all. If your goals involve a short timeframe, playing it safe might not be good enough. But for someone with more modest goals and a couple of decades to work with, the "slow and steady" approach could be smarter.

Start by taking a look at your investing goals. Can you reach your goals by gradually growing your money over a long period of time? Or do you have lofty goals that require big gains quickly?

Also remember that different portfolios can handle different levels of risk. A large and carefully diversified portfolio can usually rebound from a loss due to a risky investment, while a smaller portfolio could be destroyed by too many risks or even one big risk.

Consider, for example, two portfolios – one with $25,000 and one with $250,000. Now, let's look at how different levels of losses could affect each of them...

Amount Lost	Percent of $25,000 Portfolio	Percent of $250,000 Portfolio
$1,000	4%	0.4%
$2,500	10%	1%
$5,000	20%	2%
$10,000	40%	4%

Losing $5,000 in the $25,000 portfolio would be devastating. It would cost you one-fifth of your savings. But a $5,000 loss in the $250,000 portfolio would only be 2%.

It's a simple example. But it's an important one to keep in mind. **And it's exactly why position sizing is so important to you as an investor**.

What is position sizing? It's the idea that you put the right amount of money into your investments relative to your total portfolio size. It's a challenging concept to grasp, though, because you have to think about something that most people don't want to think about: how much you're willing to risk losing on any single investment.

A good rule of thumb is to risk no more than 4% or 5% of your entire portfolio on any one idea. As you can see in the table, if you're risking 4% of a $25,000 portfolio, you're limiting your potential loss to $1,000 on each investment. If you're risking 4% of a $250,000 portfolio, then your potential loss would be $10,000 per investment.

An experienced investor with a good track record of success and a large portfolio can afford to take a larger position size. But if you're just starting out, you likely can't do that.

Decide what your individual investing goals are. Then figure out the amount of risk you're willing to take in each investment. Use that amount to determine how much money you should put into each position.

Regardless of your goals, losing too much of your money when you're just getting started in investing is a surefire way to ruin your financial future.

That's why you also need to use...

Exit Strategies

A good investor knows losses are part of the game. If the losses are small, they don't matter.

A bad investor sees every loss as a failure. But small losses aren't failures. They are victories – victories against big losses. And big losses have to be avoided at all costs. Nobody can survive a big loss.

If there's only one secret you must learn, it's the golden rule of trading: ***Cut your losers, and let your winners ride.*** If you refuse to do these two things, you will never be successful as an investor.

It is difficult to get everything right in any given trade – valuation, sentiment, timing, position size, etc. When you've made a mistake, admit it quickly and move on. When you get everything right, treasure it. Hold on as long as possible.

When we hire new analysts at Stansberry Research, the smartest guys have the hardest time buying into these ideas. Smart guys think they're smarter than the market. And usually they are.

The problem is, you only have to be wrong once to suffer a catastrophic loss. And everyone – I mean everyone – is wrong at least once every few years. So you have to completely rule out the possibility by being a disciplined investor and cutting your losses. There is no other way.

This is easier said than done. You need to eliminate emotion from your decision-making. You need to put as much thought into planning your exit strategy as you put into the research that motivates you to buy the investment in the first place.

An **exit strategy** is a plan for how and when you will sell your investment. If you stick to your exit strategy, it can serve as a near-foolproof way to methodically cut your losses and let your winners ride. If you follow this rule, you have the best chance of outperforming the markets. If you don't, you could lose big.

There are two main types of exit strategies we advocate: a hard stop loss and a trailing stop loss. Stop losses are when you pick a price at which to sell your investment *no matter what*. If your stock hits this price target, you must sell. It takes the emotion out of the sell decision.

A **hard stop loss** is a set price at which you sell your investment to protect against a falling market. It's designed to minimize your losses. Say you buy Stock X for $30 per share. You place a hard stop at $27. That means if the stock falls sharply at any point after your purchase, you'll cap your losses at 10%. You won't be tempted to cling on, hoping the stock will turn around, only to see it plummet farther.

A **trailing stop loss** is more flexible than a hard stop. It's designed more to protect your gains than cap your losses (although it does both). When a stock's price increases, the trailing stop rises along with it.

Let's take that Stock X example again. This time, when you buy the stock at $30 per share, you immediately set a trailing stop loss of 25% – or $22.50 per share. That means that if the stock falls from your purchase point, you'll minimize your loss to just 25%.

Instead, shares climb to $40. Unlike with a hard stop, your trailing stop follows the share price higher. Your new stop is $30 (25% below $40). Even if Stock X shares fall 25%, you'll still be breakeven on your investment. If shares continue to climb to $50, your trailing stop will continue to climb, too... to $37.50. And you'll be assured a 25% gain.

Most brokers give you the option to enter your stops "into the market" using stop-loss orders or trailing-stop orders.

But keep in mind... entering your stop price into the market also leaves you vulnerable. Investors or brokers who see your stop can manipulate the share price to push you out of the position.

The safest thing to do is track your stops privately. You can do this with a simple Excel spreadsheet or with a service like TradeStops.

– Chapter 4 –

Putting It All Together

We know marijuana use is a hot-button issue for many folks. But it's not our job to judge whether it's morally right or wrong. We simply pass along the research we'd want you to share with us if our roles were reversed.

Like it or not, support for marijuana is only getting stronger...

In 1996, California became the first state to legalize the use of medicinal marijuana. In the two decades since then, more and more states have relaxed their marijuana laws.

In 2012, voters in Colorado and Washington approved the legalization of marijuana for recreational purposes. In 2018, nine states and Washington, D.C. have laws in place that allow recreational use. Another 21 states have approved the use of medical marijuana.

Overall, more than half the country allows some sort of legal marijuana. And all of this has taken place while marijuana is still technically illegal under federal law.

Despite resistance at the federal level, Americans are becoming more comfortable with legalizing marijuana use. It won't be easy to reverse the current shift in public sentiment.

Still, the legal marijuana industry is a new trend. It's rife with risk, misinformation, and hype. And it will likely endure some hiccups in the years ahead.

You don't want to get caught up in the regulatory battles. And without a crystal ball, it's difficult to predict which individual companies will be the biggest winners in the long run.

But if you do your due diligence and decide this type of investing is for you, learning from the entrepreneurs and millionaires we've profiled in *The Marijuana Industry Insider's Playbook* could help you find legitimate opportunities with the potential for extraordinary gains in the next few years – both inside and outside the stock market.

In many cases, these folks have made a fortune without touching a single gram of marijuana.

As with any new trend, there will be many people who lose money. But if you understand how this market works and how to avoid critical pitfalls along the way, you can find companies in this burgeoning industry with the potential to generate thousands of percent returns for early investors.

This trend isn't stopping. And the experts say marijuana businesses are quietly becoming one of the best-performing investments in the world.

Whatever you decide, do your own research. Draw your own conclusions. And invest only as you see fit.